WORLD BANK TECHNICAL PAPER NO. 463

# Groundwater in Rural Development

## *Facing the Challenges of Supply and Resource Sustainability*

*Stephen Foster*
*John Chilton*
*Marcus Moench*
*Franklin Cardy*
*Manuel Schiffler*

*The World Bank*
*Washington, D.C.*

First printing March 2000



ISBN: 0-8213-4703-9
ISSN: 0253-7494

Stephen Foster is assistant director of the British Geological Survey and visiting professor of hydrogeology at the University of London. John Chilton is principal hydrogeologist at the British Geological Survey. Marcus Moench is president of the Institute of Social and Environmental Transition. Franklin Cardy is senior water resources management specialist in the Africa Technical Family at the World Bank. Manuel Schiffler is an economist in the World Bank's Middle East and North Africa Region Sector Group.

**Library of Congress Cataloging-in-Publication Data has been applied for.**

# *Contents*

**Boxes**

**Figures**



**Tables**

# *Foreword*

Groundwater has been the fundamental resource underpinning the rapid provision of more reliable, better quality, low-cost water supplies for the rural population in the developing world over the past 20 years or so. Concomitantly, many nations have witnessed an enormous increase in the exploitation of groundwater for agricultural irrigation. Access to groundwater is thus a major factor enabling rural populations to achieve food security, increase their productivity and move beyond subsistence.

Whilst these developments have provided major benefits in terms of rural living standards and poverty alleviation, concerns are arising over certain issues, most notably the operational sustainability of individual water sources, the natural occurrence of groundwater of unacceptable quality in some areas, and, most importantly, widespread evidence of degradation of the resource base itself.

The preparation of this paper, which was undertaken by a team of widely experienced groundwater specialists, has been coordinated by the British Geological Survey and involved in-depth consultation with numerous World Bank staff. The work was financed by the World Bank and the (British) Department for International Development. It provides a systematic in-depth review of issues that have emerged in the 1990s and suggests the way forward towards more efficient and sustainable utilization of groundwater resources in rural development.

The target audience includes senior staff of national governments responsible for provision of rural water supply and sanitation, for promoting agricultural development and for managing land and water resources, together with the staff of the international support agencies and nongovernmental organizations charged with providing financial and technical assistance in these areas. Numerous World Bank task managers have reported they are encountering serious groundwater overdraft and pollution problems with increasing frequency, and have emphasized the lack of definitive information on effective ways to address such problems. The hope is that this paper will:

- Raise their awareness of the constraints on and threats to sustainable use of groundwater for rural development
- Provide them with a useful guide when considering new project proposals with a groundwater dimension
- Persuade them of the urgent need for increased investment and more appropriate institutional arrangements for the sustainable management of groundwater resources.

Ashok Subramanian
Senior Water Institutions Development Specialist
The World Bank

# *Abstract*

Groundwater is of major importance to rural development in many countries of the world. As a result of its widespread distribution, low development-cost and generally excellent quality, it has been the fundamental resource allowing the rapid development of improved domestic water supplies for the rural population and in many areas has also supported a major increase of highly-productive agricultural irrigation. Groundwater resources are thus vital for meeting an array of basic needs, from public health to poverty alleviation and economic development.

As a result of the high rates of abstraction required for irrigation, however, in some areas there is significant concern about sustainability of the resource base, because of falling groundwater tables and near-irreversible aquifer deterioration through saline intrusion. There are also additional sustainability concerns as a result of the increasing incidence of groundwater pollution from over-intensive or inadequately managed agricultural cultivation practices.

This paper is based on review of the evolving situation during the 1990s in a substantial number of developing nations. It aims to raise awareness of the key linkages between groundwater and rural development, and to identify appropriate technical and institutional approaches for improving the operational reliability of waterwells and the sustainability of groundwater resources as a whole. To achieve this will require recognition that hydrogeologic and socioeconomic diversity necessitates a flexibility of management response. The unifying concept of the paper is the definition of action to reduce the growth in groundwater abstraction and to constrain subsurface contaminant load, within a phased process of institutional development built upon sound technical evaluation and increasing stakeholder engagement.

# *Acknowledgments*

The potential value of producing a World Bank Technical Paper on this subject was identified by John Briscoe (World Bank—Senior Water Adviser), following an internal review of the past 10–15 years of World Bank experience with projects on rural water supply and agricultural development with a significant groundwater-related component. This review was undertaken by Stephen Foster (British Geological Survey) and Franklin Cardy (World Bank-Africa Technical Department) during April-July 1998, having been instigated by the World Bank"s Water Resources Management Thematic Group, led at that time by Ashok Subramanian. He and Andrew Macoun (of the World Bank-MENA Region), who subsequently took over as coordinator of this work, are both thanked for their personal interest and valuable inputs to the production of the paper. The work has been encouraged by the interest of John Hodges and Ian Curtis of the (British) Department for International Development—Engineering Division. The Management Committee of the Thematic Group (Geoff Spencer, Theodore Herman, and Ashok Subramanian) is to be thanked for their continued interest and support.

The authors wish to acknowledge the practical assistance of numerous World Bank staff in issue identification, data collection, policy discussion and editorial view, including Keith Pitman, Stephen Mink, John Shepherd, Lee Travers, Douglas Olson, David Grey, Salman Salman, Karin Kemper, Christopher Ward, Adel Bichara, and Ariel Dinar. The authors also wish to thank the following British Geological Survey colleagues for providing valuable data on groundwater in the context of the development of rural water supplies and agricultural irrigation: Adrian Lawrence, David Kinniburgh, Pauline Smedley, and Jeff Davies.

The first author also acknowledges valuable discussions and written contributions on the general theme of this report with Dr Hans Wolter (UN-FAO—Director of Land and Water Development Division), Professor Ramon Llamas-Madurga (Universidad Complutense of Madrid, Spain), Ing Ruben Chavez-Guillen (Comision Nacional del Agua-Gerente de Aguas Subterraneas, Mexico), and Prof K. Palanisami of Tamil Nadu Agricultural University, India. Ing Ignacio Lopez-Cortijo (UN-FAO) provided assistance in abstracting data from the AQUASTAT system.

Lastly a very special thank you to Theresa Blackwell and Gill Tyson for their major efforts in handling the presentational aspects of the document, through its various drafting stages.

# *Executive Summary*

*Sink in despair on the red parched earth,*
*and then ye may reckon what water is worth.*
*Traverse the desert and then ye can tell,*
*What treasures exist in the cool deep well.*

Elisa Cook
(poet: Southern Africa 19th Century)

*An aquifer that is almost always full,*
*is almost as badly managed,*
*as one that is almost always empty.*

David Burdon
(hydrogeologist: Ireland 20$^{th}$ Century)

The utilization of groundwater resources has facilitated the *rapid, low-cost provision of more reliable, good quality, water supplies for the rural population* across extensive areas of Asia, Africa and Latin America. While many key issues in this respect have been addressed, some persistent problems (such as improving the operational reliability of groundwater sources) and other emerging concerns (such as the hazardous or unacceptable natural quality of certain groundwaters) require systematic attention.

In many nations there has been a *major increase in the use of groundwater for agricultural irrigation.* This has not been restricted to semi-arid regions, but has also occurred in more humid areas, to provide a greater intensity, or more security, of cropping on existing cultivated land, rather than bringing new land into production. Moreover, there is increasing evidence that the use of *groundwater can be an important factor in promoting increased irrigation efficiency and water productivity.*

However, there are concerns about the operational reliability of irrigation wells. As a result of the much higher rates of abstraction required for irrigation, in some areas there is an even greater concern about the sustainability of the resource base itself, including falling groundwater tables, interference with downstream users and irreversible aquifer deterioration through saline intrusion and ground compaction. An additional issue is groundwater pollution from inadequately managed or over-intensive agricultural practices.

The principal objectives of this paper are thus:

- To highlight the major benefits of groundwater use in terms of rural well-being and income, and raise awareness of the various important (but complex) linkages between groundwater and rural development
- To provide balanced analyses of the factors influencing the reliability of individual groundwater supplies and the degradation of the overall groundwater resource
- To identify appropriate technical and institutional approaches to the challenge of improving the operational reliability of waterwells and the resource sustainability of aquifers in the context of rural development.

The organization of the paper is summarized in Figure 1, which serves as a general guide to its scope and application. It is important to appreciate that in areas where groundwater utilization is restricted to the level of domestic water supply and livestock watering, interest will be confined mainly to Chapters 1 and 2, and sometimes to Chapter 4.

Chapter 1 *(General Introduction)* details the importance of groundwater for domestic and agricultural water supply, and introduces the key linkages between rural development and groundwater resources. A hierarchy of issues and concerns is defined, which ranges from the constructional adequacy and operational reliability of groundwater sources for both domestic and irrigation water supply to resource degradation issues arising out of the development of intensive irrigated agriculture. The degree of difficulty in managing groundwater resources for rural development shows wide variation with environment and this chapter provides an introduction to hydrogeological diversity, whose appreciation is essential if the development process is to work with (rather than against) nature. It also identifies the diverse group of stakeholders in groundwater use for rural development and analyses the way in which they should be involved in the promotion, construction and operation of groundwater supply projects.

**Figure 1: General scope, organization and application of technical paper**

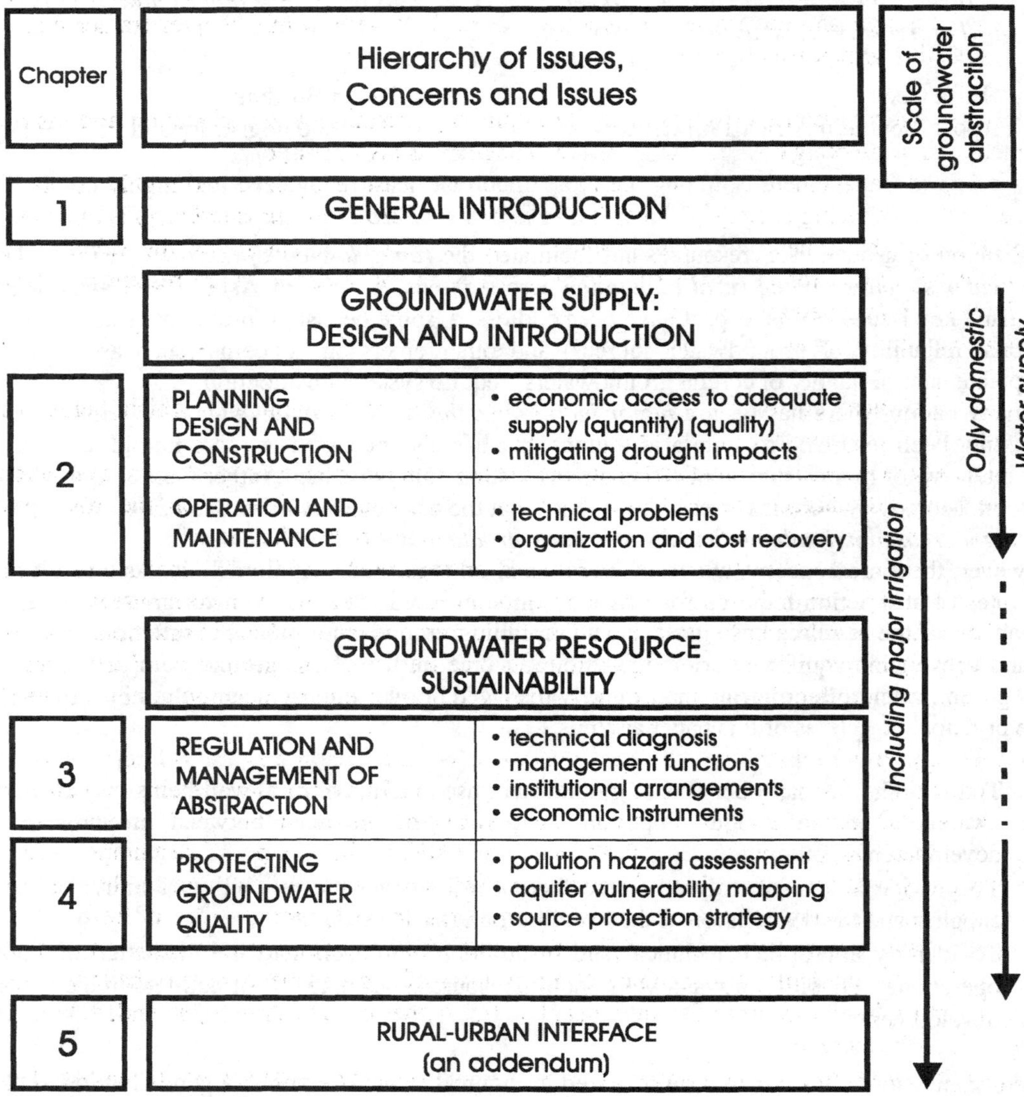


Chapter 2 *(Groundwater Supply: Design and Operation)* discusses the issues relating to the provision and operation of groundwater supplies for rural development, both at a small scale for domestic use and livestock watering, and where higher rates of water supply are of interest for piped water supply in rural towns and villages, and for intensive irrigated agriculture. Much progress was made in this context during the 1980s but there is still widespread need to ensure that:

- Well siting and design procedures benefit from being more closely correlated with aquifer
- Hydraulic structure and from systematic hydrogeological evaluation of the security of supply during extended drought
- Operational reliability of water supply sources is improved by community participation, initially through defining the required service level and subsequently through taking responsibility for both the physical and financial aspects of well maintenance
- Natural hydrogeochemical controls on groundwater quality, and the hazard of encountering unacceptable quality for potable supply are appreciated; since these act as a given, constraining the siting, design and cost of new sources, they are dealt with in this chapter (rather than later under protecting groundwater from pollution).

The key role of local water-user associations in improving irrigation-water allocation and distribution, and their potential in promoting cost-effective well maintenance, is also stressed.

Groundwater management is among the most important, least recognized and highly complex of natural resource challenges facing society. Chapter 3 (*Groundwater Resources Sustainability*) is thus the core of the paper and argues that a new approach is widely required, putting emphasis on the value of groundwater resources and the need for proactive participatory management in areas where resources are subjected to heavy demand for irrigated agriculture.

Among the key issues analyzed in detail are:

- The historical context of much groundwater resource development which helps define major obstacles that have to be overcome
- The key management functions, including the need for realistic hydrogeological evaluation of aquifer recharge, discharge and response to abstraction, strategic planning on the role, priorities and valuation of groundwater, definition and review of water rights allocation
- The promotion of effective tiered institutional arrangements and flexible management schemes, with user participation at the appropriate scale through aquifer management committees
- The potential role and limitations of economic instruments (such as abstraction charges and water markets) in groundwater management, and the need to eliminate progressively certain subsidies (especially on electrical energy for pumping) which can act as an incentive for excessive abstraction.

A critical question in the definition of many aquifer management strategies will be the optimum role for groundwater storage. In many ways the vast natural storage of groundwater systems is their most valuable strategic asset. On the one hand important components of the economic and environmental value of groundwater (such as pumping costs, individual accessibility for the poor, sustaining some freshwater wetlands and dry weather stream flow) depend on the depth to water table and not on the volume in storage. On the other hand, in many situations groundwater storage is the only major source of freshwater in extended drought, and ways need to be found to exploit this asset whilst mitigating the impacts on groundwater level related services, in particular by adequate compensation of those dependent on shallow wells for water supply. A further issue discussed in some detail is the scope and constraints on undertaking the artificial recharge of aquifer storage.

Chapter 4 (*Protecting Groundwater Quality*) summarizes the evidence of increasing degradation of groundwater quality and the threat to its potability due to leaching of nutrients and pesticides from agricultural soils, as well as the salinization of groundwater as a result of agricultural practices. The threat appears more severe and imminent in low-efficiency irrigated agriculture, and can arise regardless of the source of water supply involved. This subject, which is not always well appreciated by the agricultural development sector, is reviewed in some detail. The management response proposed is to focus much needed pollution control measures in more vulnerable aquifer recharge areas of potable groundwater sources used for piped water supply.

Overall, wide hydrogeologic and socioeconomic diversity represents a major challenge for groundwater resources management, and it is not possible to be highly prescriptive in this context. Nevertheless, *diagnostic tools can be (and have been) identified to enable resource managers and project planners to characterize the key elements of common situations and define a more sustainable way forward.* Although groundwater management and protection appear complex, the actual process of beginning to develop capabilities need not be. Furthermore, while strategies must ultimately reflect local conditions, *the overall approach to strategy development can utilize common starting points.* Where justified this might include immediate action to reduce the growth in groundwater abstraction and/or to constrain subsurface contaminant load *within a phased process of institutional development built upon sound technical evaluation, raising public awareness and increasing stakeholder involvement.* In many cases, entry points will exist in the form of specific regional concerns or local interest groups. They can be used to mobilize stakeholder participation, highlight policy issues and develop pilot activities. Proactive participatory management will represent a significant cost increment for groundwater development, but this may be a small price to pay for a secure source of reliable water supply in drought, compared to the cost of surface water supplies for irrigation.

It is recognized that the distinction between rural and urban development is somewhat arbitrary; nevertheless it is considered valid given the project focus of this paper. Urban groundwater resource management issues have been systematically treated in *World Bank Technical Paper 390 (Groundwater in Urban Development).* As a corollary *Chapter 5* deals with some special concerns about groundwater resources at the rural-urban interface and especially to three specific aspects:

- Competition for groundwater resources between agricultural irrigation users and urban water supply companies
- The fact that siting of urban groundwater sources in adjacent rural areas may lead to demands for constraint on local agriculture in the interest of protecting groundwater quality in wellfield capture areas
- The benefits that can accrue from substituting urban wastewater for local groundwater as a source of irrigation water supply, and the potential impact on water quality in aquifers that can occur if this is not adequately evaluated and planned.

# 1

# GENERAL INTRODUCTION

## Importance of Groundwater Supply in Rural Development

### *Domestic and Livestock Water Supply*

Groundwater has been the fundamental resource allowing the economical and rapid development of more reliable, improved quality, water supplies for a large proportion of the rural population across extensive areas of Asia, Africa and Latin America (Clarke and others, 1996). This crucial and formidable task gained momentum during the UN Drinking Water and Sanitation Decade of the 1980s and continues to this day. The successful development of groundwater has led to significant improvements in human health and the quality of life in innumerable village communities of Africa and Asia, in particular.

Many areas with favorable hydrogeology now have coverage of domestic waterwells for rural village populations. The major residual development challenges are:

- To tackle areas with less favorable hydrogeological conditions
- To address the need for improved maintenance and operational sustainability of systems already developed.

In the African and Latin American context, waterwells have also been of primary importance in the development of extensive livestock rearing in the semiarid regions. This aspect of agricultural development, however, has not been without its problems. In some areas there has been a tendency in wetter years to overstock in relation to land capacity during drought, resulting in subsequent heavy over-grazing and soil erosion in the vicinity of livestock-watering boreholes.

### *Agricultural Irrigation*

During the last 10 to 20 years, there has been an enormous increase in the utilization of groundwater resources for agricultural irrigation, because of their widespread distribution and low development cost (Clarke and others, 1996). Groundwater has been at the heart of the "green revolution" in agriculture across many Asian nations, and has permitted cultivation of high-value crops in various arid regions.

Groundwater has also provided security against drought in areas where irrigation with surface water resources has been deficient during dry years. Moreover, the use of groundwater can be a major factor in promoting increased irrigation water-use efficiency and agricultural water productivity. This is because the energy costs associated with pumping are often higher than for surface water, providing an incentive to increase water conservation or to irrigate high-value crops, because groundwater sources are generally far more reliable during drought and because groundwater is sediment-free, readily allowing the introduction of water-efficient irrigation technology.

Furthermore, the scale of groundwater development has facilitated tubewell operation at the level of individual farmers or small collective groups, and this has offered greater flexibility of irrigation scheduling and much simpler distribution systems, resulting in generally higher crop yields and irrigation water productivity. Moreover, it has allowed responsibility for maintenance to be devolved. Such developments, can, however, result in:

- Poor standards of irrigation well construction, which may compromise water-source reliability in unfavorable hydrogeological conditions
- The proliferation of waterwells which may lead to groundwater resource competition and storage overdraft, in situations where resources are significantly constrained by limited recharge.

*Groundwater Use Statistics*

Comprehensive statistics on the use of groundwater for agricultural irrigation are not available, but Table 1 gives an idea of its relative importance in a range of countries. One very important example is the current situation in India (World Bank, 1998). Here groundwater supplies directly about 80 percent of domestic water use in rural areas, together with more than 50 percent of that used for irrigated agriculture. The resource is thus of major importance as a source of drinking water and food security and is vital for meeting an array of basic needs from public health, poverty alleviation to economic development (Kahnert and Levin, 1993). The sustainability of the resource base is thus a critical issue in these contexts.

**Table 1: Statistics on agricultural irrigation, drainage and groundwater use for selected nations**

| *Country* | *Year* | *Irrigated area (kha)* | *Irrigation water use* ($mm^3/a$) | *Origin of water* Sw(%) | gw (%) | *Drained area (%)* |
|---|---|---|---|---|---|---|
| Bangladesh | 1993/95 | 3,750 | 12,600 | 31 | 69 | 40 |
| China | 1990/93 | 48,000 | 407,800 | 78 | 18 | 42 |
| India | 1990/93 | 50,100 | 460,000 | 41 | 53 | 12 |
| Indonesia | 1990/96 | 4,430 | 69,200 | 99 | 1 | ? |
| Malaysia | 1994/95 | 360 | 9,700 | 92 | 8 | ? |
| Nepal | 1994/95 | 1,130 | 28,700 | 74 | 12 | ? |
| Pakistan | 1990/91 | 14,330 | 150,600 | 66 | 34 | 36 |
| Mexico | 1995/97 | 5,370 | 61,200 | 63 | 27 | ? |
| Peru | 1992/95 | 1,200 | 16,300 | 89 | 11 | ? |
| Argentina | 1994/95 | 1,550 | 18,600 | 75 | 25 | ? |
| Kenya | 1990/92 | 70 | 1,570 | 99 | 1 | ? |
| South Africa | 1991/94 | 1,270 | 9,580 | 82 | 18 | ? |
| Zambia | 1992/94 | 50 | 5,320 | 95 | 5 | ? |
| Egypt | 1992/93 | 3,250 | 45,400 | 96 | 4 | 90 |
| Tunisia | 1990/91 | 310 | 2,730 | 39 | 61 | 52 |
| Morocco | 1989/91 | 1,090 | 10,180 | 69 | 31 | ? |
| Mali | 1987/89 | 80 | 1,320 | 97 | 3 | 7 |
| Jordan | 1991/93 | 60 | 740 | 40 | 55 | 6 |
| Iran | 1993/93 | 7,260 | 64,160 | 50 | 50 | 1 |
| Saudi Arabia | 1992/93 | 1,610 | 15,310 | 3 | 96 | 3 |
| Syria | 1992/93 | 640 | 13,600 | 40 | 60 | 43 |

*Note:* Although the best available, these figures do not distinguish supplementary from near-continuous irrigation, or the type and value of crops grown from different water sources, and they also do not adequately represent conjunctive, use which is known to be practiced in numerous areas.
*Source:* From UN-FAO-LWDD AquaStat database.

*Groundwater Resource Characteristics*

The characteristics of groundwater utilization for rural development are compared with those of conventional surface water schemes in Table 2. These factors explain not only the major development of groundwater resources for rural development in many nations, but also the illogical approach to groundwater resource development in some others.

The interaction of groundwater and surface water resources greatly favors their conjunctive use in irrigated agriculture, since this is capable of:

- Providing greatly increased water supply security during dry seasons and drought episodes
- enabling tail-end users in irrigation-canal command areas to improve water-service levels
- Reducing evaporation losses from surface water impoundments by allowing their storage to be exploited earlier in the dry season
- Improving drainage and reducing the possibility of rejected groundwater recharge in the wet season.

**Table 2: Comparative characteristics of groundwater and surface water resources in relation to rural development**

| *Characteristics* | *Groundwater resources and aquifers* | *Surface watercourses and reservoirs* |
|---|---|---|
| *Hydrogeological* | | |
| • Storage volumes | very large | small-to-moderate |
| • Resource areas | relatively unrestricted | restricted to watercourses and canals |
| • Flow velocities | very low | moderate-to-high for watercourses |
| • Residence times | generally decades/centuries | mainly weeks/months |
| • drought propensity | generally low | generally high |
| • evaporative losses | low and localized | high for reservoirs |
| • resource evaluation | high cost, significant uncertainty | lower cost, but still uncertainties |
| • abstraction impacts | delayed and dispersed | immediate |
| • natural quality | generally (but not always) high | very variable |
| • pollution vulnerability | variable natural protection | largely unprotected |
| • pollutant persistence | often extreme | mainly transitory |
| *Socioeconomic* | | |
| • public perception | mythical, unpredictable | aesthetic, predictable (but implies loss of valuable land) |
| • development cost | modest | very high (unless also for power generation) |
| • development risk | less than often perceived | more than often assumed |
| style of development | mixed public and private finance, individual or community operated | larger publicly financed and operated schemes |
| • project promotion time | short-to-moderate | long |
| • irrigation efficiency | frequently high | generally low |

*Note:* Various contrasting, and inadequately appreciated, features are revealed.
*Source:* Llamas, 1998)

## Effects of Agricultural Development on Groundwater

### *Recharge and Drainage Modifications*

The importation of surface water and introduction of irrigated agriculture causes major modifications to the soil moisture regime, and generally results in substantially increased infiltration (Foster and Chilton, 1998). Not all soil infiltration results in groundwater recharge to deep aquifers, but excess irrigation is a major source of groundwater recharge and under arid climatic conditions may reinitiate deep infiltration in areas where little if any has occurred in decades, centuries, or even millennia. The above also applies when local groundwater is the major source of irrigation except that in this case no net increase of groundwater resources will occur (only recirculation).

Irrigation efficiency is defined as: (water taken up by irrigated plants)/(water supplied for irrigation). Of the fraction of applied water not taken up by the irrigated crop:

- Some will be lost directly through (non-beneficial) evaporation or evapotranspiration
- Some will become surface runoff either directly or indirectly via the soil drainage system or perched water tables (together termed "irrigation return flow")
- Some will infiltrate into the unsaturated zone and recharge the main groundwater system below.

In more arid situations (and in the absence of regional aquifer flow systems) excess irrigation is likely to be the dominant component of local aquifer recharge (Foster and Chilton, 1998) (Figure 2). The corollary is that if irrigation efficiency is increased groundwater recharge decreases, but this obvious fact is often overlooked in catchment-level water management planning.

**Figure 2: General trend of groundwater recharge rates from excess rainfall and irrigation with climatic type**

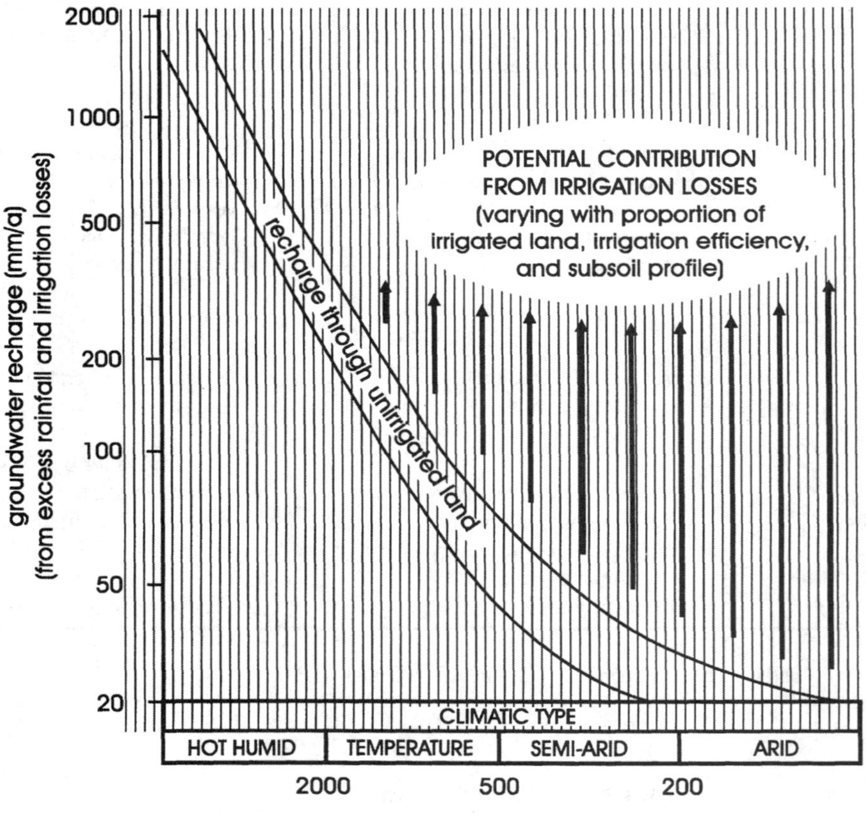


*Note:* This refers to well-drained soils overlying an unconfined aquifer and illustrates the major significance of irrigation losses for groundwater recharge in the more arid climates; the quality of this recharge, however, can in some circumstances be relatively saline.

Groundwater recharge from irrigated agriculture occurs by three distinct mechanisms:

- Directly from unlined (and in some cases lined but leaky) primary and secondary canals, and even from some agricultural drains
- Directly from irrigation water distribution systems below this level
- Through irrigation in excess of plant requirements at field level.

The potential for groundwater recharge will vary across and along irrigation areas, with higher rates from unlined canals on alluvial terraces, for example, and with groundwater discharge (rather than recharge) to the agricultural drainage systems in some low-lying areas. Where groundwater is the major (or only) source of irrigation water, the areas will normally be well drained.

In very low-lying areas, or where the soil profile is generally of low permeability (or has some low permeability horizons), rising water level or shallow perched water bodies are likely to develop. This ultimately leads to soil water-logging and salinization through direct evaporation, unless drainage is introduced to remove excess groundwater. Although this issue is outside the scope of the present paper, it should be noted that groundwater salinization caused by this process is more extensive worldwide than that resulting from saline intrusion of aquifers due to inadequate resource management.

*Quality Impacts*

The fact that an important proportion of groundwater recharge in many areas originates as infiltration on agricultural land (especially where irrigation is practiced) also has a negative side - namely the risk of excessive leaching of nutrients and pesticides (Foster and Chilton, 1998). A close correlation between agricultural development and groundwater quality in underlying shallow phreatic aquifers is widely observed (Figure 3).

**Figure 3: Correlation between land use and groundwater nitrate concentrations**

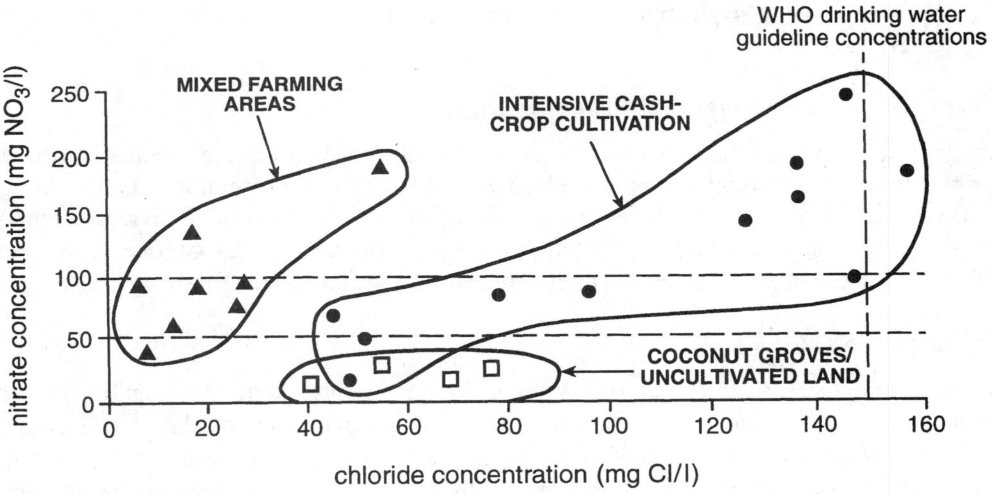


*Note:* The data shown refer to a thin shallow coastal limestone aquifer in northwestern Sri Lanka that reacts quickly to land-use change; the conversion of land to intensive irrigated "cash crop" cultivation has clearly had a major impact on groundwater quality

In practice, the rates of leaching will vary widely with cropping regime, soil type and hydrogeological conditions (aquifer vulnerability), with irrigation water efficiency and the continuity of crop coverage being especially critical factors. In certain monocultures on permeable soil profiles, especially those involving soil ploughing and fallow periods, the leaching losses may be severe.

The principal impact is on the potability of groundwater for rural water supply, at farm, village and small town level. It is rare that the level of contamination is such that it can prejudice the use of groundwater for agricultural irrigation itself, except in a few cases of exceptionally severe nutrient and/or pesticide leaching adjacent to an area of cultivation of highly sensitive crops.

Transpiration of water by plants concentrates dissolved salts in the root zone, and periodically there may be considerable leaching of salts from irrigated agricultural soils. In extreme cases where major groundwater recirculation occurs, salt fractionation can cause a troublesome quality impact. The situation is further aggravated where excess irrigation gives rise to leaching of salts held in arid zone soil profiles. Such processes may be just as widespread as the problems of saline intrusion due to overabstraction of groundwater, but less commonly recognized. Thus in areas of major development of irrigated agriculture from groundwater in arid climates, it is important to evaluate both the water and the salt balance.

## Key Groundwater Development and Management Issues

The process of identifying key issues can be usefully initiated from a "development project focus" by adopting the following subdivision:

- Internal factors within projects (those that can be controlled by the project and which determine its cost effectiveness and operational reliability)
- External impacts of projects (side effects caused by projects on third parties and the environment).

Beyond the scope of individual projects is a range of emerging resource management issues at aquifer level which require a much broader approach. Thus overall, three major groups of issues have been identified. These are analyzed in detail sequentially in the chapters that follow, but the underlying concepts involved are introduced briefly below.

### *Economical Access and Operational Reliability of Supply*

Hydrogeological factors are dominant in determining whether a groundwater source can be constructed at tolerable cost to provide a supply of initially adequate yield, acceptable natural quality and drought reliability. Operational reliability relates to the longer-term sustainability of yield for the individual groundwater source (as opposed to total yield of the aquifer system) and is influenced by its design, operation and maintenance, together with adequate financial resources and administrative arrangements for this purpose.

### *Aquifer Depletion-Related Effects*

This concept recognizes that all aquifer systems are to varying degree susceptible to such effects as interference between production wells, diminution of groundwater discharge affecting downstream riverflows, freshwater wetlands or brackish water lagoons, the encroachment of saline water through lateral intrusion or up-coning, and in certain cases land subsidence. These may threaten the sustainability of the resource base itself (Reisner and Bates, 1990) and the agricultural food production dependent upon it, although current estimates of the potential impact are not based on sound concepts or data (Postel, 1999). Sooner or later, and to varying degrees, groundwater abstraction needs to be controlled to avoid or mitigate the more serious of these effects.

*Diffuse Groundwater Pollution*

The development of agriculture, whether rainfed or irrigated (and regardless of water source), can result in excessive leaching of agrochemicals to aquifers and lead to long-term deterioration of groundwater quality (in relation to use for potable water supply), especially where intensive monocultures are sustained through large applications of fertilizers and pesticides. The issues of aquifer pollution control and groundwater source protection thus need to be addressed.

## Variation of Issues with Scale of Groundwater Exploitation

The question of scale of groundwater exploitation is important, since the extent to which projects have external impacts will vary widely between the extremes:

- Small-scale domestic and livestock water supply
- Large-scale agricultural irrigation schemes.

Within this range, there was traditionally a distinction between small-scale garden cultivation and large-scale (institutionally promoted) irrigation schemes. In reality today there is a near-continuum between the two, with much successful groundwater irrigation occurring at the intermediate scale of multi small-well development since this allows:

- Private (individual or group) operation of each well, avoiding some of the past problems of centralized operation
- Small water distribution networks, avoiding the high leakage losses of many larger schemes.

The most logical subdivision of groundwater development scale is thus now that summarized in figure 4, which is based essentially on the distinction between manual and motorized pumping.

Resource overexploitation problems relate mainly to groundwater supply for agricultural irrigation, since in the case of domestic/livestock water supplies resource sustainability issues are only significant for shallow (low-storage) aquifers in arid regions during extreme drought.

## Variation of Issues with Hydrogeological Regime

The (natural) hydrogeological environment exerts the dominant control over the availability of groundwater resources for any type of rural development and the corresponding water supply development costs and difficulties. Geodiversity, in general, and hydrogeological variability in particular, are still poorly appreciated by many working in water/land resource management and in promoting rural development projects. There is need that they recognize intrinsic constraints on groundwater development, and try to work with nature rather than against it, when identifying and promoting groundwater development schemes for the benefit of the rural community.

Moreover, hydrogeological setting influences the scale of potential side effects of large-scale land development for agricultural cultivation since it determines:

- the susceptibility of groundwater resource exploitation to negative consequences
- the vulnerability of groundwater resources to pollution from agricultural land-use practices.

**Figure 4: Variation of well yields and abstraction requirements for different types of rural groundwater use**

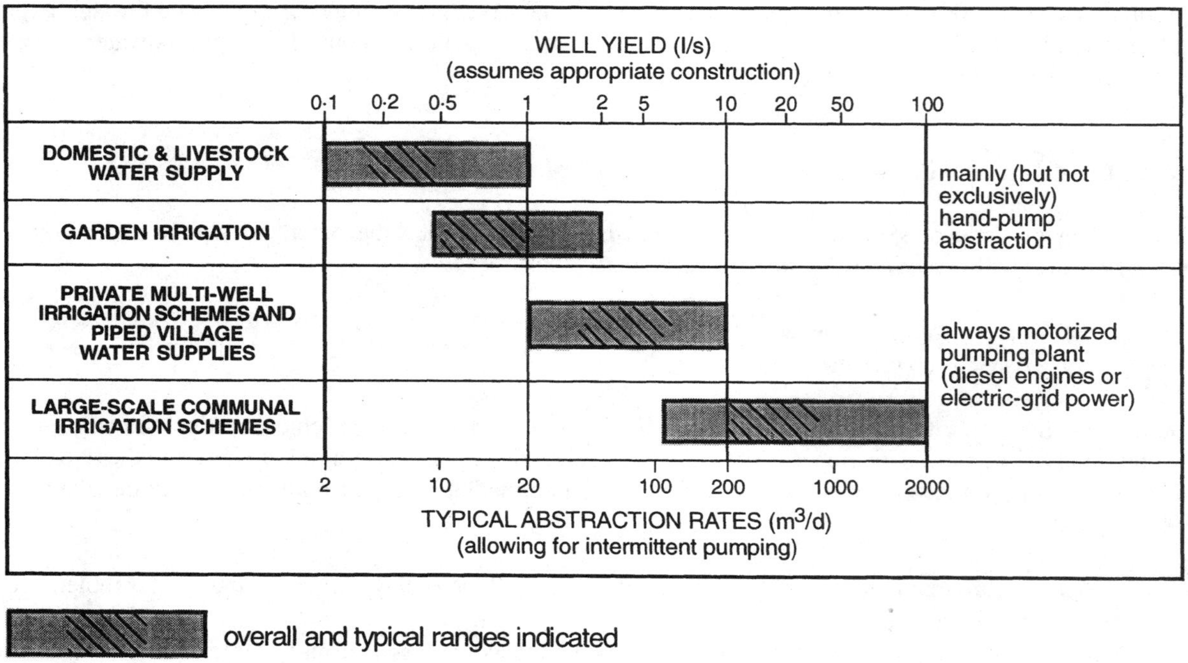


*Note:* Those which do not require motorized pumping plant do not threaten groundwater resource sustainability and thus need only minimal regulation, appropriate hydrogeological investigation and engineering design protocols

A general indication of the occurrence and flow of groundwater in major regional aquifers, and its variation with climatic regime, is given in box 1 (Foster, 1993). A highly simplified classification and description of the more common hydrogeological environments in the developing world is given as table 3. In reality, certain other factors must also be considered, such as the degree of aquifer confinement, the prevailing climatic regime, the constraints on groundwater recharge, and the natural aquifer discharge.

The hydrogeological environment imposes constraints on the access to groundwater for rural development. Such constraints can be absolute in terms of large-scale groundwater development for irrigation in certain environments. There is a clear correlation between groundwater supply development costs and hydrogeological complexity (that is decreasing hydrogeological predictability), which varies widely with hydrogeological environment; figure 5 gives an indication of the relative position.

## Analysis of Issues from Stakeholder Perspectives

From an early stage in rural development projects, it is important that the full range of stakeholders is specifically identified, and their interests in groundwater sketched out in a general way. The actual and potential role of these various stakeholders in groundwater project development, system operation and/or maintenance, and even in resource/environmental management, is a recurrent and developing theme of this paper.

## BOX 1 : Groundwater Occurrence and Flow

• All freshwater found underground must have had a source of **recharge**. This is normally rainfall, but can also sometimes be seepage from rivers, lakes or canals. Infiltrating water accumulates above an impermeable bed **(aquiclude)** forming an underground reservoir **(aquifer)**. The strata above the aquifer water table, through which vertical infiltration occurs are termed the **vadose (or unsaturated) zone**. Aquifers tend to fill up until water reaches the land surface, where it flows from the ground as springs or seepages at some locations, the discharge providing the dry-weather flow (or baseflow) of lowland rivers. The aquifer becomes saturated to a level where outflow matches recharge. From the management viewpoint, note that most continuous groundwater abstraction, for consumptive use in (or export from) the catchment, will have some impact on dry-weather riverflows, the discharge of captured springs and/or groundwater levels in wetlands.

• The aquifer flow regime, storage capacity and yield productivity depend upon the hydraulic characteristics of the porous and/or fractured media involved, and vary widely with the geology.

• Groundwater systems are dynamic with groundwater continuously in slow motion from zones of recharge to areas of discharge. Tens, hundreds or even thousands of years may elapse, since flow rates do not normally exceed a few metres per day and can be as low as a metre per year. It will thus be apparent that a surface contamination incident will normally take a long time to affect deep water-supply boreholes, a fact which has major policy implications for pollution control.

• Aquifers in recharge areas are generally **unconfined** but elsewhere, and normally at greater depths, groundwater is often found to be **confined** by virtually impermeable layers. In this instance, when wells are first drilled, water is encountered under pressure and rises on its own, sometimes even to the ground surface. The piezometric head/surface is the level to which the water from a given aquifer will rise. In some cases, the overlying strata are less permeable but do not completely prevent the vertical passage of water, and the aquifer is then said to be semi-confined, below an aquitard. Such **semi-confined aquifers** can still receive vertical recharge, but at much lower rates, which will be significant in terms of the long-term sustainability of groundwater abstraction.

**(b) semi-arid regions**

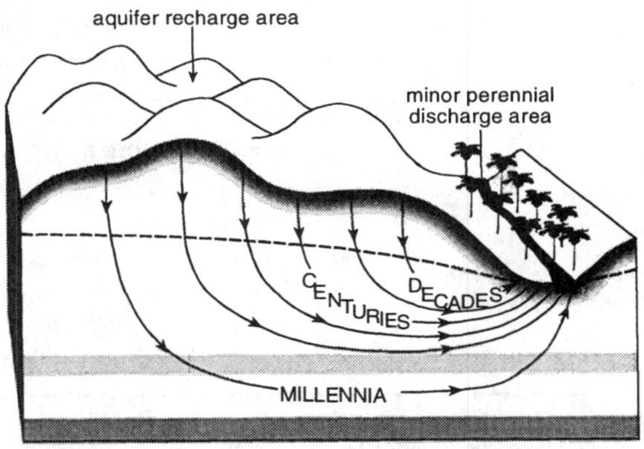


***Inset :*** *Typical groundwater flow systems (Foster & Hirata, 1988)*

**(a) humid regions**

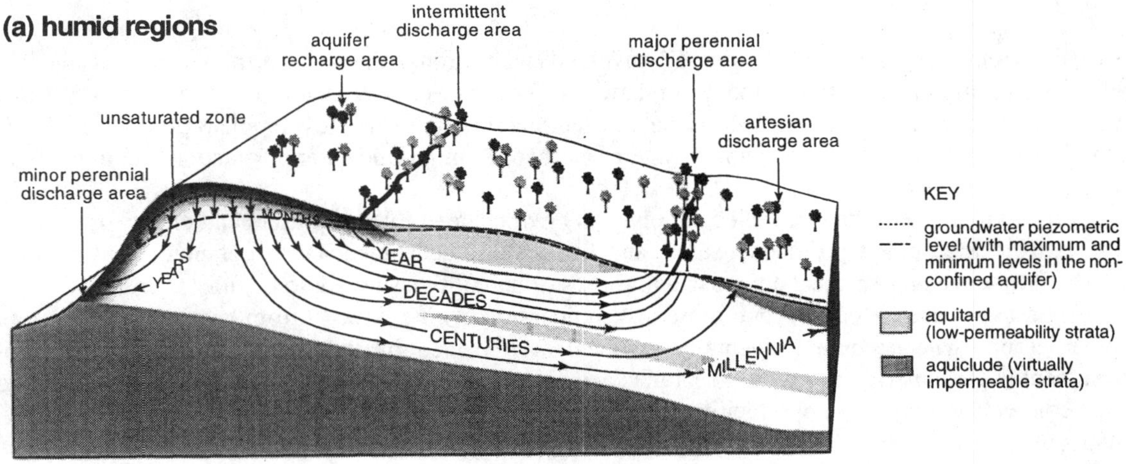

**Table 3: Characteristics of principal hydrogeological systems**

| *Hydro-geological environment* | *Type of deposits* | *Mode of formation* | *Distribution and thickness* |
|---|---|---|---|
| *Major Alluvial Formations*<br>(a) inland<br>(b) coastal<br>(MAF) | Gravels, sands, silts and clays | unconsolidated detritus deposited in riverbeds and deltas, primary porosity/permeability usually high | both areally extensive and of significant thickness |
| *Inter-Montane Basins*<br>(a) colluvial<br>(b) volcanic<br>(IMB) | pebbles, gravels, sands and clays; sometimes with lavas and pyroclastics | formed by in-filling of faulted troughs in mountain regions and can include lake deposits; recent lavas and pyroclasts also usually highly porous (but older volcanic deposits more consolidated) | less extensive than most alluvial and coastal plain sediments but can be very thick |
| *Consolidated Sedimentary Aquifers (CSA)* | (a) sandstones (sometimes also stratiform basalts) | compacted marine or continental deposits; degree of consolidation increases with depth/age and reduces primary porosity/permeability but with significant fracturing | difficult to generalize, but can form extensive aquifers of substantial thickness |
| | (b) limestones | derived from shell fragments/reef detritus; compacted and often with karstic fissures enlarged by solution | difficult to generalize, but can form extensive aquifers of substantial thickness |
| *Recent Coastal Limestones (RCL)* | limestones and calcareous sands | coral limestones and skeletal detritus often only loosely cemented; porosity/ permeability very high | limited extension, fringing coastlines or islands |
| *Weathered Crystalline Basement (WCB)* | grading from weathered rock to residual clays | deep weathering of igneous/metamorphic rocks usually producing mantle of moderate porosity/low permeability | very extensive, but aquifers of small capacity and normally restricted to upper 20 m or less |

*Note:* The five broad groups of aquifers commonly occurring in tropical latitudes of the developing world are shown; the Major Alluvial Formations and Weathered Crystalline Basement are by far the most extensive in geographical distribution.

The main groups of stakeholders directly involved with groundwater in rural areas, and the normal (traditional) timing of their involvement in relation to project evolution is given in Figure 6. The timescale can be from 1 to 5 decades, as a result of the considerable inertia of the development process, coupled with the large storage/slow response of aquifer systems to changes in groundwater abstraction and in contaminant load.

A complexity of interests is revealed and there is obvious need for water users and energy suppliers to be involved from the project promotion stage, and for development agencies to continue their involvement throughout (not just up to project commissioning), if sustainability issues are to be fully addressed.

It is necessary also to look beyond those stakeholders benefiting directly from groundwater development to other groups who become incidentally involved or impacted by the activity (Figure 6). The role of government is particularly difficult to generalize and present, but the trend (which needs to be encouraged) is for governments to play more of a facilitating role than a developmental one. However, in reality some parts of national and provincial government inevitably will be involved with the promotion and construction side of development, while other arms of the same government are involved in resource regulation and environmental protection. These activities can come into conflict.

The scale and timing of benefits and disbenefits to the various stakeholder groups are also indicated in a general way. It is important that the perspective of these stakeholder groups on groundwater, in terms of resource accessibility, ownership, limitations, linkages and externalities is fully appreciated. There is also a need for public awareness to bring the various groups of stakeholders on to a "common playing field" so that they can participate more equally in groundwater project development, and to develop a consensus among them for action on groundwater resource management.

**Figure 5: Variation of groundwater supply development options/costs with aquifer type**

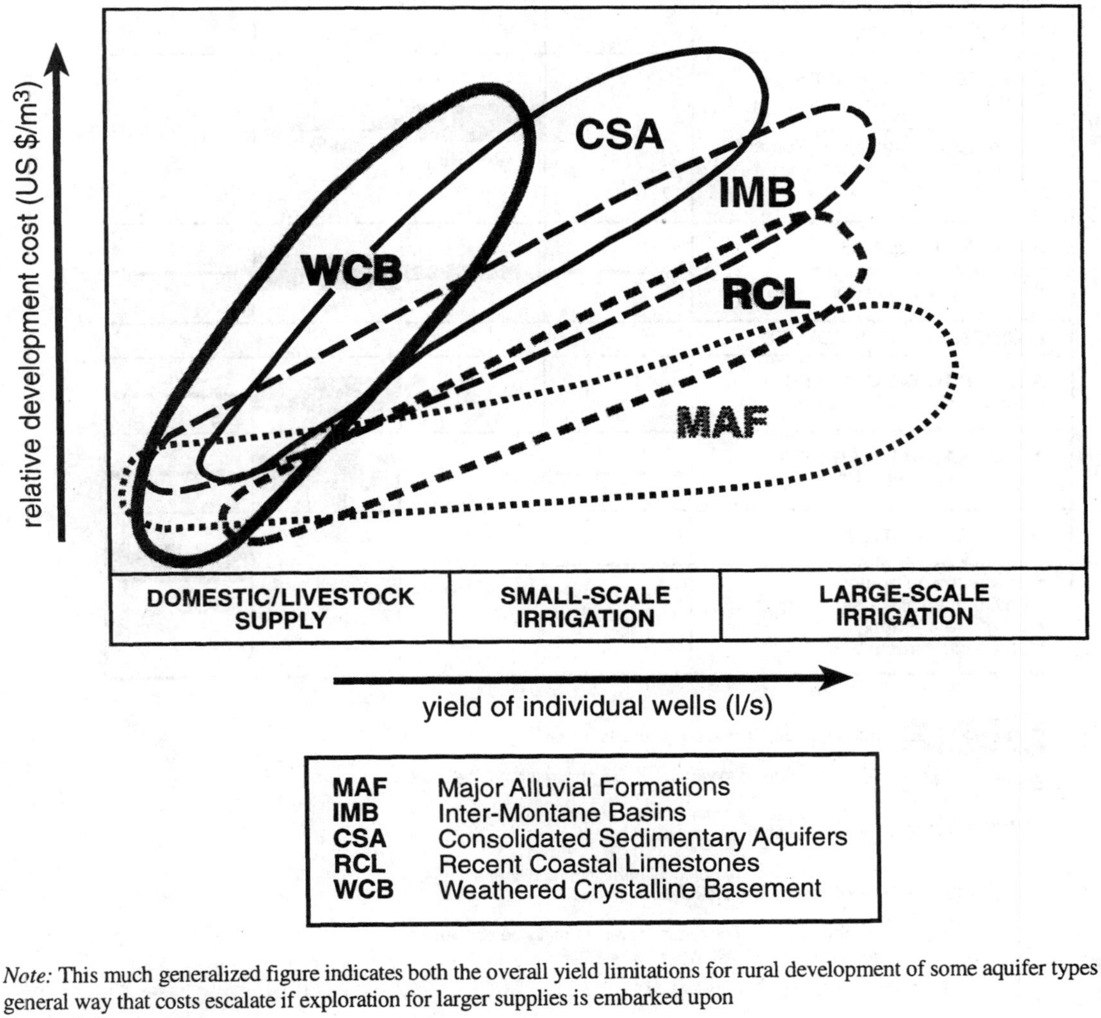


*Note:* This much generalized figure indicates both the overall yield limitations for rural development of some aquifer types and the general way that costs escalate if exploration for larger supplies is embarked upon

**Figure 6: Analysis of actual and required stakeholder participation in rural groundwater development for agricultural irrigation**

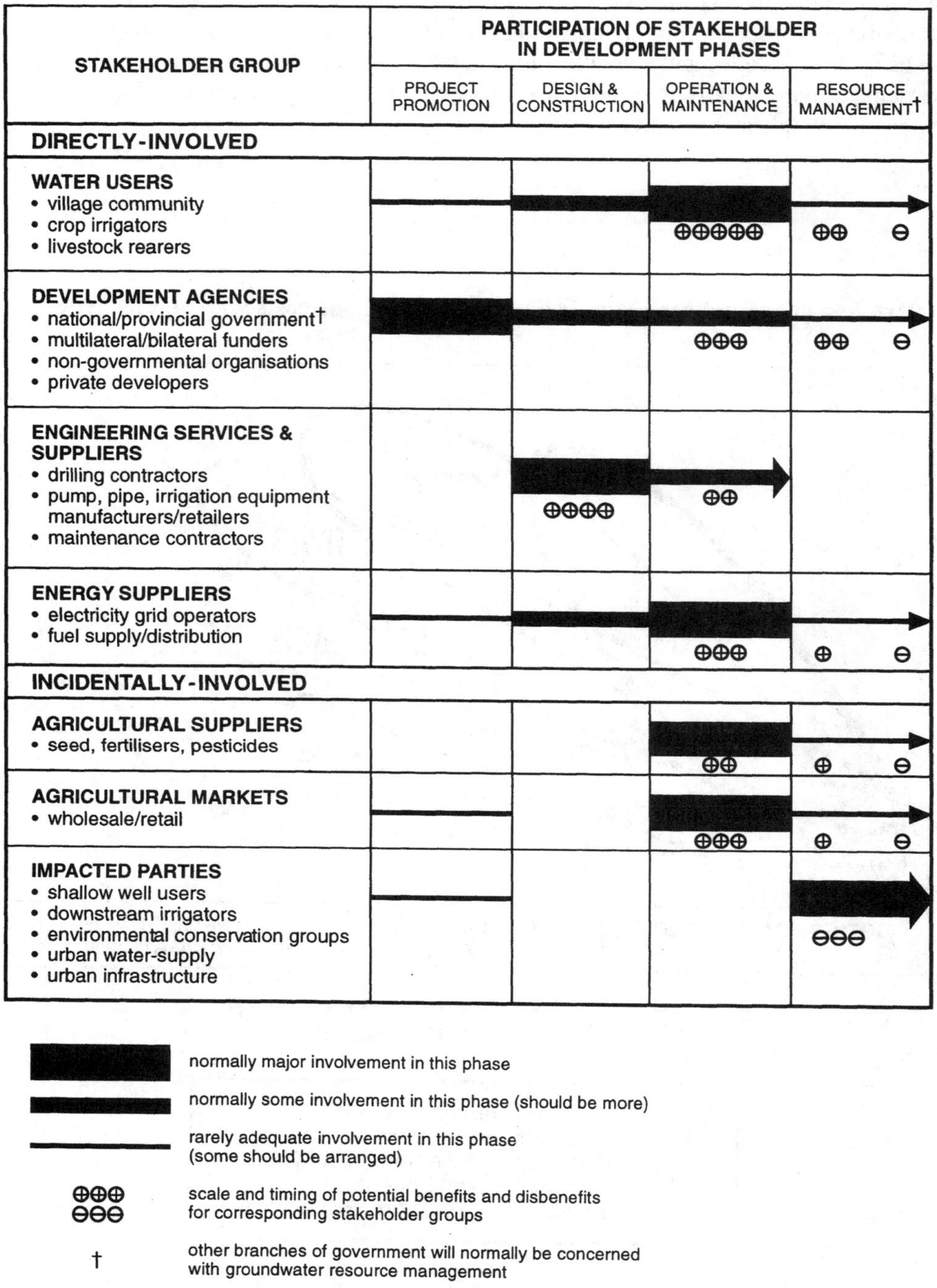


*Note:* The horizontal time-scale may be from 1 to 5 decades; both the typical current situation and the preferred approach is indicated; the corresponding picture for domestic water supply development is less complex but stakeholder participation is equally necessary.

# 2

# *Groundwater Supply: Installation and Operation*

## Context of Main Issues

During the International Drinking Water Supply and Sanitation Decade (1980s) rural water supply coverage increased from 30 to 63 percent and the population with basic services decreased from 1613 to 989 million, despite the large growth in rural population during the period (Subramanian and others, 1997). The program depended heavily upon groundwater to meet the demand for domestic water supply in rural areas, rapidly and economically.

Although much has been achieved, the need to meet the demand from steadily expanding populations remains a major challenge. Moreover, the sustainability of established supplies, through adequate operation and maintenance, was recognized as a key issue. Full consultation and participation of the "beneficiary community" is also now regarded as essential, and the importance of cost recovery, adequate maintenance and source protection are seen as critical to operational sustainability.

The international NGOs led the way in helping communities develop the improvements in water supply they wanted. While these organizations were regarded with suspicion by some governments, much progress was made in demonstrating the effectiveness of community participation in the development, operation and maintenance of rural domestic water supply wells.

Large-scale groundwater resource development for irrigated agriculture has a relatively short history. The possibilities for groundwater exploitation changed radically with advances in the turbine pump, deep drilling technology and geological knowledge, notably from the mid 1960s in Pakistan and more widely in the 1970s. Groundwater development itself was often carried out on an individual, or small-scale cooperative, basis without the parallel development of an effective institutional framework for water provision. Hence there is now a considerable challenge to maintain groundwater supplies operationally and to promote sustainable use of groundwater resources as a whole.

In order to meet the expanding demand for rural water supply, certain *hydrogeological factors* are critical to well siting and design. The nature of groundwater occurrence and broad range of hydrogeological environments has been summarized in Box 1 and Table 3 respectively. The ability of aquifers to store and transmit (or yield) water exhibits substantial variation from place to place and not all the defined hydrogeological environments can meet the needs of all users (Figure 5). If the global range of climatic regimes is superimposed on hydrogeological setting, then the complexity of intrinsic constraints on groundwater development is accentuated. Some environments will provide only modest yields, and then only if the most favorable locations can be selected. Others will provide moderate-to-high yields from almost anywhere in the aquifer, provided the well is correctly constructed.

More specifically, where there is adequate surface water, crystalline basement rocks are often not considered to be aquifers, but if rainfall is lower and surface water scarce or intermittent, then the crystalline basement can form the only economically exploitable source of rural water supply.

## Access to Adequate Water Supply

*Domestic and Livestock Well Construction*

During the International Drinking Water supply and Sanitation Decade much effort was concentrated on improving rural water supply in Sub-Saharan Africa and South Asia. Although, with a few notable exceptions (Arlosoroff and others, 1987; Wright and Burgess, 1992; Van Dongen and Woodhouse, 1994), the technical experience gained has not been systematically reported and disseminated, knowledge of the corresponding hydrogeological environments has still increased substantially. Thus, for example, the position of the crystalline basement rocks in Figure 5, invariably having low-to-moderate yield potential and small volumes of storage, has been confirmed by extensive bodies of field data (Box 2), collected in rural water supply projects (Chilton and Foster, 1995).

***Siting and Design Criteria.*** Exploration for groundwater has been a key task for geologists for many decades. Early practitioners generally used electrical resistivity geophysics in simple standard ways, having been assigned to (rather than trained in) groundwater exploration. More recently, with the greater availability of trained hydrogeologists and the extension of projects into more difficult terrain, a wider range of siting techniques and better interpretation have been employed.

*Groundwater exploration* should be phased, employing increasing levels of sophistication. Five successive levels of investigation can be defined:

- Inventories of existing geological, hydrogeological and borehole data
- Remote sensing using satellite imagery and aerial photographs
- Reconnaissance hydrogeological fieldwork (including geomorphological characterization and examination of existing water supply sources)
- Surface geophysical surveying by various techniques (according to their cost and to local conditions)
- Detailed hydrostratigraphic survey, including exploratory drilling and pump testing.

Each successive level adds more detailed information concerning hydrogeological conditions.

In reviewing well siting approaches in Africa, Van Dongen and Woodhouse (1994) found that geophysical surveys were often employed where the first three phases listed above were only cursorily performed. If this happens, very useful and inexpensive information is neglected, increasing the overall cost of siting. Further, each *geophysical site survey* is often treated as an entirely individual task. While the survey operator may have some degree of accumulated local knowledge of the relationship between geophysical soundings and hydrogeological conditions, usually no systematic use is made of the body of existing data. Moreover, lack of communication between those responsible for siting and construction (for various reasons) may mean that there is also no proper feedback from the actual drilling results to improve the operation and interpretation of future geophysical siting surveys.

Although significant geophysical effort may be put into the borehole site selection process, success even in terms of modest hand-pump yields is not guaranteed. Comparative studies of the "success" of different approaches to borehole site selection are complicated by the differing definitions of success. It may not be easy to evaluate rigorously the benefits of a specific methodology because "with" and "without" technique performance data rarely exist in comparable environments, such that other factors can be eliminated. There is thus little "control" in the scientific sense and the best than can be hoped for is a comparison between figures before and after a certain method was employed. This can make the selection of the most suitable approach difficult, and the choice of geophysical technique in particular is too often made without proper regard for the hydrogeological environment (Table 4).

## BOX 2 : *Village Water Supplies from the Weathered Crystalline Basement in Sub-Saharan Africa*

• Crystalline basement rocks underlie much of Sub-Saharan Africa, and their mantle of weathering products forms a shallow but low-productivity aquifer. This provides a vital source of water-supply for the rural population and their livestock who inhabit these areas.

• The ancient land surface of the region has been exposed to prolonged weathering which has formed a mantle of alteration products, known as the regolith. This can be up to 30 or 40 m thick, and comprises the residual soil and underlying weathered (disaggregated and sometimes clay-rich) crystalline rock. The physical and chemical processes of weathering have produced dissolution and leaching of the less stable minerals, leading to increases in porosity and permeability. The transition to fresh, unweathered rock is usually gradual, and the basal part of the weathering sequence is likely to be the most permeable and productive.

• Remarkably consistent hydrogeological conditions have been revealed from detailed investigations for rural water-supply provision in western, eastern and southern Africa, and experience gained there is also applicable to similar environments elsewhere.

• The ability of the regolith to provide adequate yields for rural water-supply or small-scale irrigation depends on the available saturated thickness. This is in turn a function of climate; shallow water tables are more likely in the less arid parts of the region where recharge is greater. Where the regolith is thick and the water-table shallow, yields may sustain a handpump, but where the weathering sequence is much thinner and the water-table deeper, the regolith is unlikely to be usable.

• Experience from projects in Malawi (Inset I) suggest that 10-12 m of weathered material below the water-table is sufficient, providing appropriate borehole designs are selected (Chilton & Foster, 1995). While the mean values suggest a rather consistent and uniform aquifer, in practice groundwater conditions vary greatly over short distances depending on local hydrogeological and geomorphological factors.

| PARAMETER | LIVULEZI | | LILONGWE | | DOWA WEST | |
|---|---|---|---|---|---|---|
| | no. | mean | no. | mean | no. | mean |
| borehole depth (m) | 145 | 23.1 | 212 | 17.9 | 103 | 25.6 |
| regolith thickness (m) | 134 | 21.5 | 101 | 12.0 | 25 | 25.1 |
| water struck (m) | 75 | 10.2 | 191 | 8.6 | 95 | 14.4 |
| saturated regolith (m) | 80 | 10.9 | 192 | 9.9 | 97 | 11.5 |
| water level (m) | 139 | 7.3 | 185 | 5.7 | 97 | 10.1 |
| borehole yield (l/s) | 139 | 0.75 | 187 | 0.73 | 94 | 0.43 |

***Inset I:*** *Characteristics of weathered crystalline basement aquifer in Malawi rural water-supply projects*

• The most favourable locations are often associated with geological features (such as fault zones and fractures) which encourage deeper weathering, and may also be local focii of recharge. This variability is illustrated by Inset II. For this reason, considerable effort is required in site selection to determine locations with thick saturated regolith. Even so, the potential for achieving higher yields for small reticulated supplies is not great, although the use of collector wells can help to maximise the productivity of the shallow regolith aquifer.

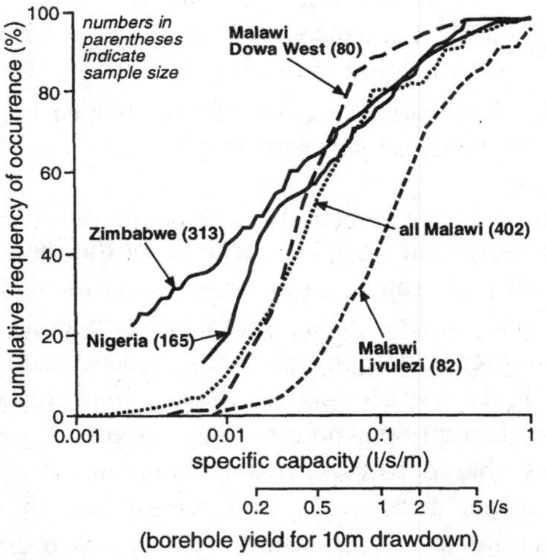


***Inset II:*** *Cumulative distributions of specific capacity for boreholes in weathered crystalline basement aquifers in Zimbabwe, Nigeria and Malawi. (most can achieve the minimum of 0.2 l/s for handpumps, but few 1.0 l/s for motorised pumps).*

**KEY ISSUES:**

- **the water-table depth and the thickness of weathered basement should dictate the approach to siting, design and construction of water-wells.**
- **the development of larger supplies requires greater investment to locate the most favourable zones of higher transmissivity and maximum available drawdown**

In the national rural water supply program in India, from the late 1960s to the late 1990s some 2.8-3.0 million hand-pump boreholes were constructed, of which perhaps 80 percent are in the more difficult hydrogeological environments (weathered crystalline basement of the "peninsula states" and basaltic volcanic rocks of the Deccan plateau). In spite of investment in geophysical equipment and training, and the gradual adoption in some states of additional aids such as aerial photographs and remote sensing, the overall "success rate", measured against the target hand-pump yield of 0.2 l/s appears to have increased only from 75-80 percent to 85-90 percent. Although site selection procedures have improved, this has been partly offset by progressive movement into more difficult areas as rural water supply in India approaches full coverage.

**Table 4: Suitability of geophysical methods in different hydrogeological environments**

| *Hydrogeological environment* | *Electrical resistivity* | | *Seismic refraction* | *Electro-magnetics* |
|---|---|---|---|---|
| | *Sounding* | *Profiling* | | |
| Major Alluvial Formations | ++ | + | + | o |
| Consolidated Sedimentary Aquifers | + | + | + | o |
| Volcanic Formations | + | + | o | + |
| Weathered Crystalline Basement (regolith) | ++ | + | ++ | + |
| Weathered Crystalline Basement (fractured bedrock) | + | ++ | ++ | ++ |
| Fresh/Salt Water Interfaces | ++ | + | o | + |

++ very suitable
+ suitable
o not suitable
*Note:* This gives a general overview of the more commonly used techniques.
*Source:* Van Dongen and Woodhouse, 1994.

Site selection must take account of the views of the community who will use the supply. Experience has shown that a strong sense of *community ownership* is required from the technology choice and site selection stage, if the users of the supply are to operate and maintain it effectively. However, this will tend to slow down the construction process, and hence add to overall cost. In the World Bank Swajal Project in Uttar Pradesh, the use of local NGOs as support organizations to promote community involvement (including the choice of supply technology) has increased the per capita well cost far above the national average, but results suggest that construction standards are also higher. In early programs, in which social mobilization components were weak, sites were often chosen far from communities or such that certain people had preferential access and others were excluded. Exclusion has been such a serious and prolonged issue that the latest guidelines from the Rajiv Gandhi National Drinking-Water Mission for the provision of rural water supply specifically allocate disproportionate funding to redressing previous anomalies.

Finally, the precise choice of site, after hydrogeological and community criteria have been satisfied, must take account of risks associated with local sources of pollution, flooding and erosion, physical accessibility for construction and future development in the neighborhood.

One outcome of the extensive programs in Africa of the 1980s and 1990s was the increased attention given by hydrogeologists to the weathered crystalline basement. While reference has already been made to the impact this had on siting techniques, attention was also turned to applying sounder design principles. Traditionally, boreholes were drilled through the weathered zone (regolith) deep into the unweathered rock below in the hope of finding water-yielding fractures. The weathered regolith was cased out, and the fresh rock left as open hole, resulting in inefficient boreholes with high hydraulic losses allowing entry of fine materials. As the potential of the regolith zone became apparent it was evident that the relative position of the water table and base of the regolith should dictate well siting and design (Figure 7). Another outcome was the realization that the choice of drilling equipment was a key issue in relation to cost and success, and that in many cases simpler technology was more appropriate (Table 5).

**Table 5: General summary of drilling methods and constraints for waterwell construction**

| *Drilling equipment* | *Hand digging* | *Hand-operated rig* | *Cable-tool Rig* | *Small air flush rig* | *Multipurpose rotary Rig* |
|---|---|---|---|---|---|
| Capital cost approx (US$K) | 1 | 1-5 | 20-100 | 100-250 | 200-500 |
| Running cost | very low | low | low | medium | very high |
| Training needs | very low | low | low-medium | medium | very high |
| Repair skills | very low | low | low-medium | medium | very high |
| Back-up support | very low | low | low-medium | medium | very high |
| Range of penetration rates (m/8-hr day) | 0.1-2.0 m | 1-15 m | 1-15 m | 20-100 m | 20-100 m |
| 200 mm holes to 15 m unconsolidated formation | slow | fast | fast | impossible | very fast |
| 200 mm holes to 50 m unconsolidated formation | generally impossible | slow and difficult | fairly fast | impossible | very fast |
| 100 mm holes to 15-50 m consolidated formation | extremely slow | impossible | very slow | very fast | very fast |

*Note:* The very fast rates of drilling which are possible with more sophisticated drilling machines can only be sustained if careful attention is paid to planning their logistic support.
*Source:* Arlosoroff and others, 1987.

***Financial and Economic Considerations.*** The relationship between hydrogeological factors and overall waterwell costs is given in Figure 7. For the crystalline basement rocks, costs can rise relatively rapidly but with little chance of achieving large yields. Consolidated sedimentary aquifers will generally have relatively high costs, due to deep drilling and high pumping lifts. Other formations are more unpredictable due to greater lithological variability and depth range; within these the lowest overall costs are associated with shallow water table situations, which have low drilling costs and small pumping lifts.

Actual costs would provide a clearer picture, but it is difficult to find country or region comparative figures because of differing labor costs and differences over what is included in the costing, and (of course) because such figures are rarely published. The cost data from a questionnaire survey of rural water supply projects in the late 1980s are shown in Table 6. The exploration costs (at up to 10 percent) include significant effort in remote sensing, aerial photo-interpretation and geophysical survey.

**Figure 7: Harmonizing design of rural water supply wells with hydrogeological conditions in weathered basement aquifers**

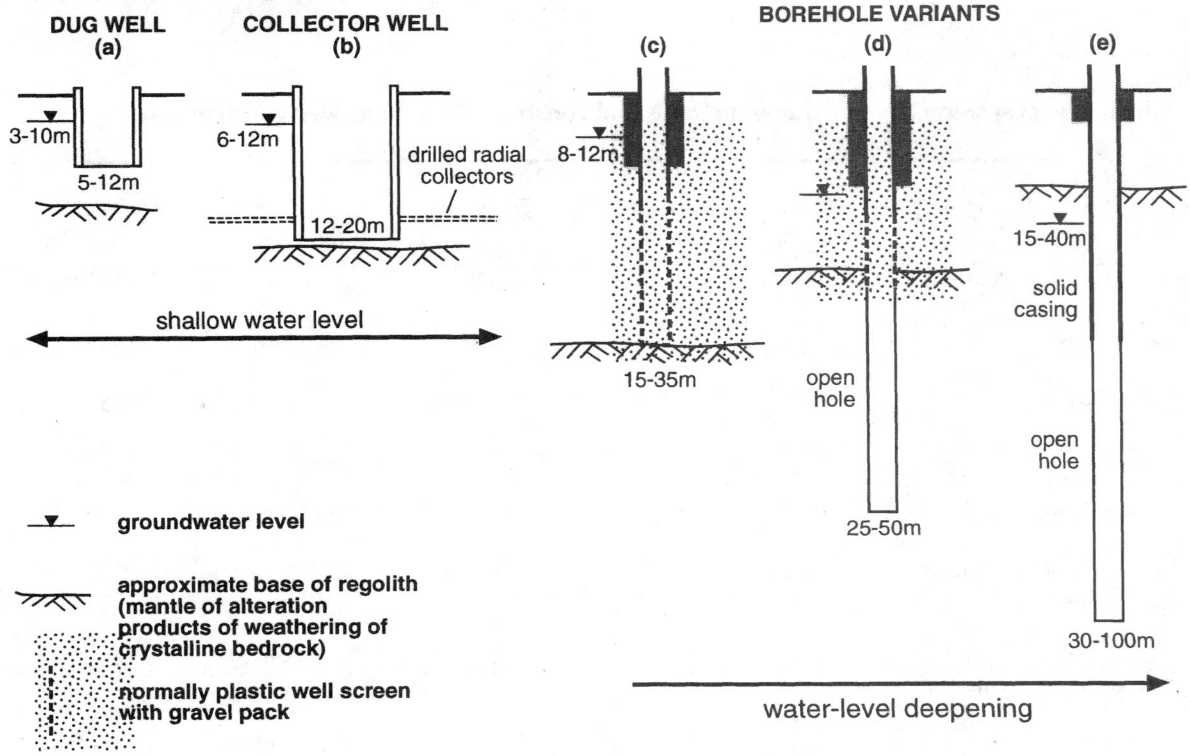


*Note:* Shallow groundwater and thick regolith (c)(d) permit simple approaches to both design and siting, but a deeper water table with thin saturated regolith (e) requires complex and costly siting of boreholes aimed at locating bedrock fractures and thin saturated regolith (a)(b) requires careful siting and increased well diameter

*Well design requirements and construction costs* are a function of:

- Depth, since the first usable groundwater may be very deep in some consolidated sedimentary and volcanic aquifers
- Diameter, since this increases significantly if a gravel pack is necessary
- Construction materials, including the need for non-corrosive casing and screen.

In the major alluvial formations high construction costs may be incurred due to the need for high-quality screens with gravel packs (especially for large yield requirements), which require large diameters and also high cleaning costs.

**Table 6: Average costs of rural water supply wells in weathered crystalline basement regions**

| *Region* | *Average cost per well (US$)* | *Number of Projects* | *Number of wells* | *Investigation costs per site (US$)* |
|---|---|---|---|---|
| West Africa | 12,000 | 12 | 6,921 | 1,200 |
| East Africa | 10,000 | 8 | 7,969 | 420 |
| Southern Africa | 2,800 | 6 | 2,751 | 210 |

*Source:* Van Dongen and Woodhouse, 1994.

The total amount to be invested in groundwater for rural development, be it for domestic supply or for livestock watering, is inevitably strongly influenced by the political process and the cost and reliability of alternative water supply sources. However, a somewhat more rigorous process can be defined to help decide how much of this should be used for a given groundwater search or well-siting method. This is illustrated for livestock watering in Botswana (Box 3), but a similar approach would be valid for domestic water supply. The analysis depends heavily upon having usable data from national archives on the success rate for so-called "wildcat drilling" (without systematic scientific investigation or siting procedures).

*Garden Irrigation and Village Systems*

As domestic water supply coverage increases, the aspirations of rural communities broaden to include provision for supplementary garden irrigation or small piped distribution systems to deliver water to individual house connections or village standpoints. There is also a requirement for small piped systems to serve rural hospitals, clinics, schools, trading and administrative centers.

Increasing awareness of the range of rural water supply options potentially available usually also implies a greater willingness to contribute towards the cost of operation and maintenance. However, the capital cost of construction is still widely regarded as a "government responsibility." Responding to these changes, implementing agencies have begun to offer communities a technology choice, in which higher levels of service require greater contributions from the users to both capital and operating costs.

The *additional yield required* to sustain garden irrigation or small reticulation systems (Figure 5) is not generally problematic for some hydrogeological environments and little supplementary effort may be required for site selection to achieve modest yield increases. Further, because the water will be reticulated to users, the need to site wells very close to demand reduces and the target area for siting can often be enlarged, which increases the chances of finding adequate yield. Site selection, however, will need to take account of topography in relation to pumping lifts and distribution tanks, and of local sources of pollution.

The higher yields required will affect construction approaches and costs. The most obvious is that motorized pumps will require boreholes of somewhat greater internal diameter. For modest yields of 2-5 l/s, diameters will typically be 150 mm compared to 100 mm (sufficient to accommodate a hand-pump). The incremental cost, however, is generally modest in terms of drilling, especially where open-hole completion is feasible. Some hand-pump programs routinely construct boreholes of sufficient diameter to allow motorized pump installation on higher yielding boreholes if required.

In alluvial formations, extra costs may be incurred for well screens and gravel packs and consequent larger diameter. In all cases, additional effort will be needed to clean (develop) the boreholes until sand-free water is discharged and to test the yield-drawdown relationships to determine optimum pump capacity and setting. Thus, while a short-duration test of 60-100 minutes may be sufficient for handpump boreholes, 12- or 24-hour tests are generally recommended for motorized pump installations.

The situation is much less favorable in the crystalline basement areas (Figure 5). A relatively modest increase in yield to 1 l/s is difficult to achieve, even with much greater effort on siting. There are numerous examples of small piped schemes based on inadequately yielding boreholes, which produce water only intermittently and making very ineffective use of the capital investment in reticulation. Van Dongen and Woodhouse (1994) found few data from the siting of higher-yielding boreholes, but estimated the costs to be around US$2000 (2-10 times greater than those for hand-pump boreholes).

Considering the much greater investment in pumps, tanks and piping for small reticulation schemes, substantially greater investigation costs can be justified if the prospect of adequate well yield results.

In Malawi, Kenya and Zimbabwe villagers have been encouraged to use the wastewater drainage from handpump aprons for vegetable plots or fruit trees. More recently this approach (Box 4) has been expanded to include the collector well concept (Figure 7), so as to better exploit the weathered basement aquifer for garden irrigation.

## *BOX 3 : Cost Effectiveness of Groundwater Exploration for Livestock-Watering Boreholes in the Botswana Kalahari*

• The rearing of livestock is the traditional enterprise in many of the world's semi-arid regions. Large tracts of land are normally grazed at low densities, and in many cases there is interest in extending and upgrading grazing areas to increase beef cattle herds. In this connection, provision of adequate water-supply represents a major (if not the largest) capital investment.

• The scale of projected water demand and the economic value of the water resource should dictate the depth of knowledge of the local groundwater system required. For livestock alone, water demands are small and dispersed (exploitation level rarely exceeds equivalent of 0.2 mm/a) and there is little risk of cattle watering depleting available groundwater, even where no replenishment is occurring. Only a qualitative knowledge of groundwater occurrence is necessary, but locating any formation reliably yielding adequate quantity (0.5l/s) and acceptable quality (TDS less than 5,000 ppm) can still pose a significant problem.

• An area of some 3,000 $km^2$ in southern Botswana was chosen for study (Farr et al, 1982) and a statistical analysis of previous waterwell drilling experience (from the national borehole archive) revealed sparse hydrogeological knowledge and difficult groundwater conditions, and more specifically that:

* the success rate where boreholes encountered Stormberg Basalt alone was only 58%, compared to 86% in other geological formations, despite the fact that boreholes were normally drilled to at least 150m
* however, where adequate groundwater was encountered in the Stormberg Basalt, it was invariably struck at shallower depths (20-40m) than in other formations (60-80m).
* groundwater occurrence at the base of the superficial Kalahari Beds was surprisingly frequent (43%) but supplies were often saline.

• The success rate that any exploration technique will enjoy is not known, but the minimum success rate (S min) necessary for it to be economically viable is:

$$S\ min = \frac{d\ So}{(d - C\ So)}$$

where d is the borehole drilling cost, So the wildcat drilling success rate and C the cost per borehole of a given technique.

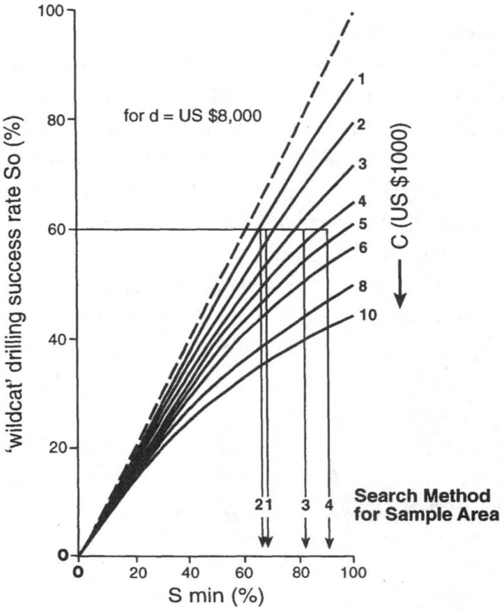


***Inset :*** *Minimum justifiable borehole success rate for given groundwater search cost*

• It can be seen (Inset) that for an So of 60% and d of US$8,000 (small contractor with slow equipment) techniques (1) Grid-Controlled Drilling and (2)-Geology-Controlled Drilling would only require an increase in success rate of 6-8% to be justified, whereas techniques (3) Long-Traverse Geophysics and (4) Short and Long-Traverse Geophysics would need much larger increases and (5) Full Hydrostratigraphic Exploration is not viable.

**KEY ISSUES:**

- **should government subsidise those initiating the groundwater search in any given area, recovering revenue from levies on late arrivals who benefit from the improved hydrogeological knowledge**
- **should government tax cattle sales more generally to finance contributions to the groundwater search and other essential ranching services**

Hydrogeological factors affecting the choice of abstraction between hand-pump and motorized-pump need to be set in their broader institutional and social context. Even where the hydrogeology is favorable, other factors such as reliability of power supplies and maintenance arrangements for more complex systems determine the success and sustainability of supplies. It is now common in southern Indian states to see traditional sources being used at different times in villages whose piped supplies operate discontinuously, negating the health benefits they were intended to impart. The intermittent nature of rural electricity supplies is a major factor, together with lack of defined maintenance responsibilities for these schemes.

*Design and Construction of Larger-Scale Irrigation Boreholes*

While groundwater has become increasingly important for agricultural irrigation, in some instances problems arise in relation to:

- Access to adequate borehole-yield at realistic capital cost
- Poor well efficiency and useful life, as a result of inadequate design and/or
- Maintenance.

Nevertheless, groundwater development (including the necessary hydrogeological studies) will be much more economic in terms of capital development costs than new surface water irrigation schemes.

The above issues show wide variation with *hydrogeological environment* (which affects the availability of supplies even for the most efficient boreholes and the difficulty and cost of design of acceptable boreholes) and (interactively) the scale of irrigation water development (which in turn influences utilization, ownership, operation and maintenance).

Only some hydrogeological environments are potentially able to provide the yield potential for larger-scale irrigation, and for piped water supply schemes in bigger villages or small towns (Figure 5). These include (among others):

- The large alluvial formations of the Indian sub-continent, South-East Asia and China
- The recent karstic limestones of parts of China and the Jaffna Peninsula in Sri Lanka
- The consolidated sedimentary formations of the Arabian peninsula and North Africa
- The intermontane basin aquifers of central Mexico.

Unfortunately, in Sub-Saharan Africa, where the need for greater crop productivity is probably greatest, such aquifers are not widely developed and the emphasis there must be on rational development of the limited potential for small-scale irrigation (Box 4).

In most favorable environments, hydrogeological criteria are not a major constraint in site selection. The most important hydrogeological consideration is that of borehole design for maximum pumping efficiency, minimum operating costs, low maintenance costs and long operating life. The minimum length and diameter of open or screened borehole required for a certain design yield can usually be estimated. For all hydrogeological environments, the depths and thicknesses of the main aquifers are required for selecting the positions and lengths of the screened sections. In the unconsolidated alluvial sediments, adequate knowledge of the grain-size distribution is also essential for selection of screen-slot opening and gravel-pack size. Important local siting considerations are likely to include topography to command sufficient land of suitable quality, land ownership and availability of power.

Assembling adequate knowledge of these characteristics requires broader groundwater studies of a more comprehensive nature than the siting investigations referred to previously. These normally comprise:

- Investigation drilling and sub-surface sampling, with down-hole borehole logging to determine aquifer geometries and characteristics
- Grain-size analysis of aquifer material and detailed test pumping to determine hydraulic properties
- Hydrochemical sampling and analyses to evaluate any groundwater quality constraints.

### *BOX 4 : Developing Small-Scale Garden Irrigation Using Collector Wells in Zimbabwe*

• The widespread uptake of small-scale garden irrigation in semi-arid areas is constrained by the availability of water. The weathered superficial materials overlying crystalline basement areas in Africa and Asia provide a thin but often extensive aquifer. Abstraction of groundwater from this aquifer for rural domestic water-supply has been by traditional, hand-dug wells where the water-table is shallow, or by deeper boreholes into the fractured rock below. The collector well offers an alternative mode of groundwater withdrawal, offering potentially larger yields.

• A collector well is a shallow large diameter hand-dug well from which horizontal or inclined boreholes, (usually four in number), are drilled from the base up to 30 m laterally into the aquifer. Research on the hydrogeology of the weathered material or regolith (Chilton & Foster, 1995) has shown that the base of the profile is often the most permeable. Constructing a dug well of 2 m diameter to the base of the weathered material allows lateral boreholes to be drilled into this more permeable zone.

• Collector wells can be used to optimise groundwater abstraction from the regolith aquifer. Higher yields and smaller drawdowns are obtained compared to conventional boreholes, utilising the storage within the well. Mean safe yields of 8 collector wells in Zimbabwe and 20 in Sri Lanka were 2.7 l/s, with drawdowns of 2-3 m, compared to typical yields of 0.1-0.7 l/s for drawdowns of up to 30 m in small boreholes; the construction of the lateral boreholes increasing the effective well radius. Comparison of yields before and after lateral drilling show improvements in the general range of 50% to 250%, with the occasional much higher figure reflecting the interception of a fracture by one or more radial.

• As collector wells are typically 10-18 m deep with the water table at 5-10 m, handpumps are well suited for groundwater pumping. The large diameter provides substantial storage which is replenished as water levels recover in non-pumping periods, and also permits the installation of more than one handpump. This ensures that the required pumping capacity can be achieved, and reduces the risk of water shortages at critical times if a pump breaks down.

• Between 1992 and 1995, the technical, economic and social factors determining the feasibility of small-scale garden irrigation using collector wells were examined (BGS-IH-MLWDZ, 1996). Collector wells were constructed at 9 locations in south-eastern Zimbabwe with unreliable rainfall in the range of 400-800 mm/a. The yield increase from lateral drilling averaged 38%, and at all sites a consistently-adequate supply of water (averaging 26 $m^3$/d) was obtained for domestic use and small-scale irrigation.

• Collector well performance was monitored from completion. This confirms that the peak water requirement to irrigate a garden of 0.5 ha and satisfy local domestic needs is 14-15 $m^3$/d. Moreover, the wells were shown to provide reliable supplies even in drier years of low recharge, and the earliest such well and garden constructed in 1991 sustained irrigation through the 1991-92 drought, albeit to a reduced area.

• Detailed socio-economic monitoring and evaluation showed the schemes to be economically viable, with an average IRR of 19%. The gardens provide major opportunities for the marketing of vegetables to surrounding areas in the dry season, and more reliable supplies of fresh vegetables to the families of scheme members. However, experience showed that implementation was not simple, and many local community and social factors influenced scheme performance.

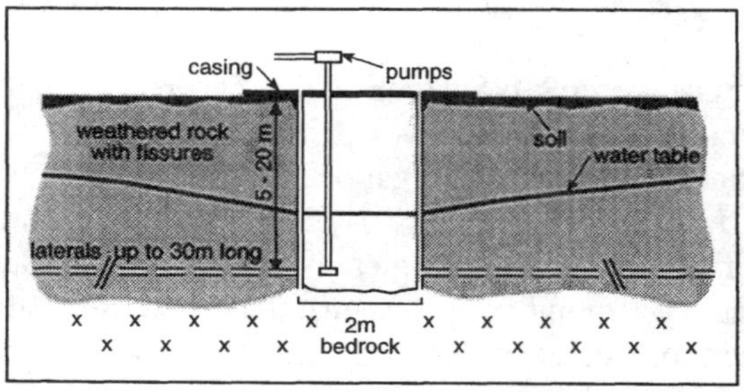


***Inset :** Schematic diagram of a collector well used for small-scale irrigation*

**KEY ISSUES**

- **although the technique has been proven effective, can construction costs be reduced further by the use of more local materials and by economies of scale in larger programmes**
- **can robust institutional arrangements be established which allow more productive water use and contribute to the cost of operation and maintenance**

Such studies are costly but, if short-cuts are taken in their implementation or interpretation, technical design flaws can lead to subsequent problems, which increase maintenance costs or shorten useful life (Driscoll, 1986) (Box 5). In the worst cases resulting borehole failure produces a major loss of investment and/or requires costly rehabilitation.

It has become more widely recognized that *borehole ownership* at the level of individual farmers is often preferable, since it offers greater flexibility of irrigation scheduling, much simpler command areas and, most importantly, devolves responsibility for well operation and maintenance to those with the greatest interest in reliable water supplies. However, this can also lead to poorer-quality and lower-cost irrigation well construction which (for reasons outlined above) may compromise water source sustainability, especially in unfavorable groundwater conditions.

*Drought Security Concerns*

A major advantage of groundwater as a source of supply arises from the "buffering effect" of aquifer storage in relation to climatic variability and changing demand, which (especially for irrigation) are often closely linked. Indeed, it is likely that the greater drought security of most wells compared to surface water sources is a major factor in explaining the normally far better economic productivity of land irrigated with groundwater.

While in most hydrogeological environments drought security will thus not be a significant concern, in certain situations the storage capacity is more limited and some wells may dry up altogether. A "groundwater drought" may result, and absolute water shortages then replace crop failure as the most critical issue for the affected population. Such was the case in Zimbabwe when after several dry years during 1988-92, normally reliable wells began to dry up. The emergency drilling programs established by government in response were at best of debatable effectiveness, since boreholes were poorly sited, community participation was negligible, construction was inadequately supervised and maintenance was not planned (Benson and Clay, 1998). As in many previous examples, stretching back to the Bihar drought of 1967, this resulted in the number of unsustainable rural water supplies increasing substantially.

For rural domestic water supply, even in relatively densely populated areas, usage amounts to an equivalent of only 1-3 mm/a of replenishment (Chilton and Foster, 1995), and localized depletion in the immediate vicinity of the well (not regional aquifer dewatering) is likely to be the usual cause of any problem. This may occur in low-permeability aquifers, such as the crystalline basement. Increasing demands of humans and livestock can mean that groundwater cannot move quickly enough and a steep dewatered cone of depression forms around the source; for this reason it is necessary to make the distinction between source and resource constraints on groundwater availability (Calow and others, 1997). The greater stress on groundwater sources during drought (as demand and pumping lift both increase) can also lead to seizure of pump bearings. Pump maintenance may become a lower priority during drought, and the result may be gradually increasing abstraction at fewer and fewer sources.

An important aspect of groundwater behavior in drought conditions is the time-lag between recharge and response in groundwater levels and well yields, in contrast to the much more rapid response of surface waters. Because of the buffering effects referred to previously, meteorological drought may not always lead to groundwater drought. Where it does, several successive years of low rainfall may be required, and the response may even not become fully apparent until after the meteorological drought has ended by return of adequate rainfall.

The *time-lag and severity of drought* impact on groundwater depend on:

- The duration of the drought episode
- The type, design and siting of groundwater supplies (shallow dug wells are likely to be more sensitive and interfluve sites are likely to be more affected than those in valleys).
- The demand on sources (as described above)
- The characteristics of the aquifer, in particular its storage (the crystalline basement and volcanic aquifers are likely to be more sensitive (Figure 6) than large alluvial formations).

## BOX 5 : Diagnosis of Borehole Deterioration and Rehabilitation Needs in the Indus Alluvial Basin of Pakistan

• Since the mid 1960s some 13,000 public boreholes have been installed by the Pakistan Water & Power Development Authority (WAPDA) in the alluvial deposits of the Lower Indus valley. These are used to provide water for irrigation and to control the water-table to prevent waterlogging and salinisation under a series of Salinity Control & Reclamation Projects (SCARPs).

• The Lower Indus valley is underlain by a sequence of fine and medium sands up to 500 m thick, containing some thin lenticular silts and clays. This constitutes a vast, unconfined aquifer with high storage and permeability characteristics, which is suitable for the abstraction of groundwater from boreholes yielding up to 150 l/s.

• Monitoring programmes established by WAPDA have shown that the yield-drawdown characteristics of many of these boreholes declined rapidly, even though the designs were apparently based on sound construction principles (Ahmad, 1990). Some have failed completely, with collapsed screens or total silting-up, while some maintained their operating efficiency. The overall result was that prediction of borehole life, proved problematic.

• Initially a standard borehole design was adopted based on reverse-circulation drilling at 600 mm diameter, mild-steel casing and slotted screen surrounded by a gravel pack. This design was shown by test pumping to be hydraulically efficient. Boreholes were equipped with turbine pumps driven by vertical shaft from electric motors, providing trouble-free operation.

• Investigation of early problems indicated that the grading of the standardised gravel pack was coarser than that required for the aquifer in some areas. Within the overall design, account needed to be taken of local aquifer grain-size characteristics, and modifications were made accordingly. Rapid loss of efficiency was ascribed to encrustation and/or corrosion of screens, and fibreglass or stainless-steel screen materials were used to combat this.

Even with improved gravel packs design performance deterioration was still encountered and it appeared that some must be occurring away from the screen, (perhaps at the interface between the gravel pack and the aquifer) or in the aquifer itself (Ahmad, 1990). One cause of decline might be damage to the aquifer fabric by re-alignment of the mica flakes in the formation close to the borehole. Hydrochemical processes are important, particularly as many of the boreholes either penetrate into saline water or produce upconing of saline water. Encrustation may occur at the gravel pack/aquifer interface even when it is not obvious at the screen itself, and chemical alteration of the mica particles in response to increasing salinity may cause expansion and permeability reduction.

• Understanding the causes of deterioration is critical for assessing rehabilitation prospects and techniques. Early attempts to restore boreholes suffering from encrustation of mild-steel screens by acidising and blasting produced only partial short-term improvements. Rehabilitation using acidising, chlorine treatment for iron bacteria, and surging/ jetting was somewhat more successful (Inset).

| PROJECT | BOREHOLES | AVERAGE SPECIFIC CAPACITY (l/s/m) | | |
|---|---|---|---|---|
| | | original | before treatment | after treatment |
| Mona | 7 | 23.6 | 10.4 | 24.0 |
| SCARP III (Alipur District) | 15 | 22.3 | 8.2 | 10.4 |
| WAPDA Construction Works | 7 | 28.5 | 2.7 | 17.1 |
| Irrigation & Power Department | 20 | 27.8 | 11.9 | 14.4 |

*Inset : Effectiveness of production borehole rehabilitation in Indus alluvial deposits at pilot level*

**KEY ISSUES:**

- **lack of investigation of aquifer site characteristics can lead to inappropriate well design and construction, and to significant performance deterioration**
- **causes of deterioration are likely to be complex, but must be adequately diagnosed if rehabilitation methods are to be selected correctly and implemented effectively**

The link between hydrogeological environment and *drought propensity* is further illustrated by Figure 8, which is based on experience of recent examples. In India 82 percent of the land area is composed of "hard-rock" aquifers (mostly crystalline basement and the Deccan Basalts) and over the last 30 years most Indian states have experienced drought conditions at different times which were severe enough to impact on groundwater resource availability. In 1967, Bihar was the focus of extreme drought and relief agencies responded by deploying drilling rigs by airfreight; this was followed by less intense droughts in 1983 and 1987. Even today, groundwater resources in the rain-shadow areas of Maharashtra (on the Deccan Basalt) are unreliable and thousands of so-called tanker villages still receive drinking water by road for up to 120 days each year. Severe African droughts in Ethiopia, Somalia, Uganda, Kenya and Sudan, while worsened by political upheaval, owe their severity at least in part to the low storage of the underlying shallow and/or thin aquifers in the crystalline basement rocks. The southern African drought of 1989-92, also largely in crystalline basement areas, had severe impacts on populations in Malawi and South Africa, as well as in Zimbabwe, Mozambique and Botswana.

**Figure 8: Variation of borehole yield predictability and drought security with principal hydrogeological environments for domestic and livestock waterwells**

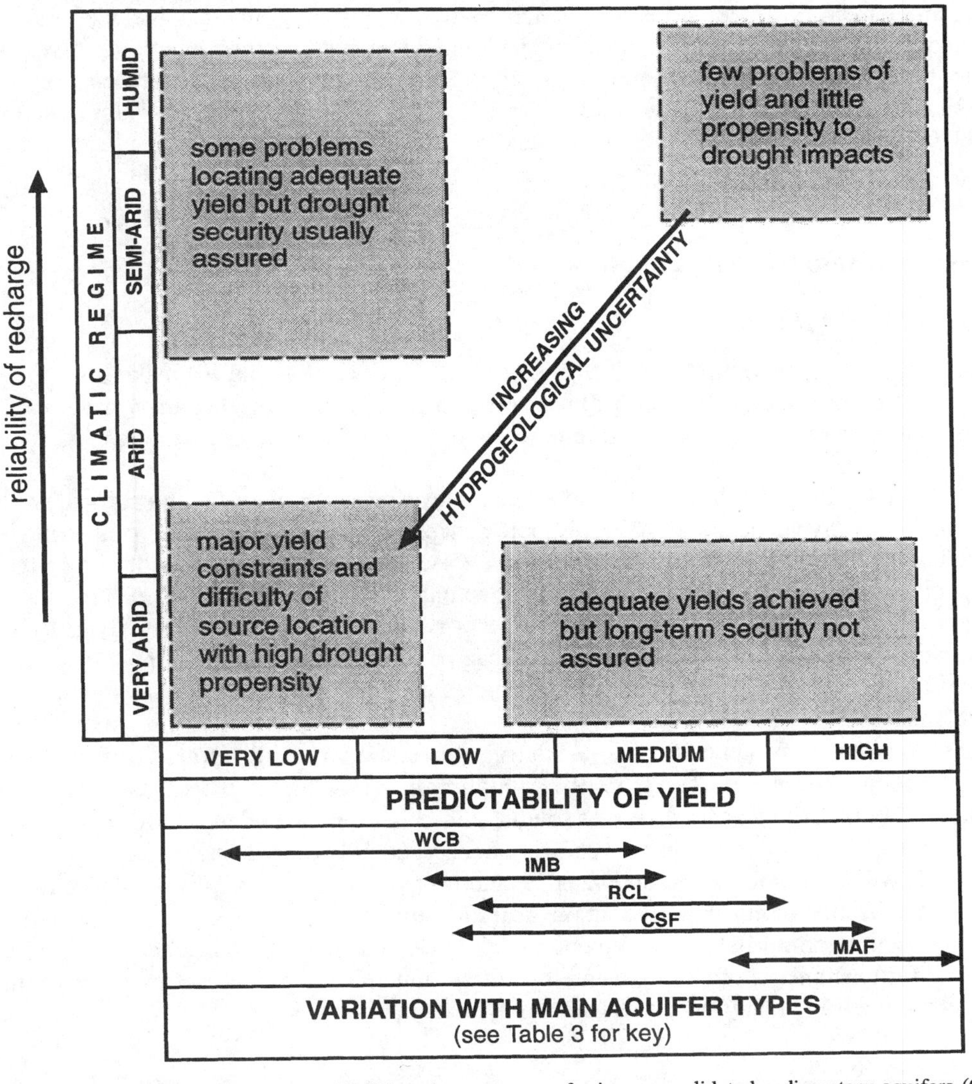


*Note:* The weathered crystalline basement (WCB) and some types of minor consolidated sedimentary aquifers (CSA) show the highest propensity to drought impacts.

A useful strategy for addressing the issue of groundwater droughts is the preparation of "drought vulnerability maps" (Calow and others, 1997) (perhaps better termed "*drought propensity maps*" to avoid confusion with other uses of the word "vulnerability"). Because drought propensity can be predicted, maps combining the factors of physical vulnerability to drought (aquifer recharge, permeability and storage) can be combined with the factors of human vulnerability (supply coverage and population demand) to produce an overall map. Thus, in areas of difficult hydrogeology (with dependence on traditional sources), and where supply coverage is low but population density high, drought impact could be most severe. Such maps could be used for:

- Warning of impending drought impacts on groundwater resources
- Allocating scarce resources in the most sensitive areas in pre-drought periods
- Making appropriate technology choices (for example, between dug wells and boreholes)
- Ensuring adequate approaches to siting and construction (particularly depth) are made.

A problem is that all too often monitoring of groundwater levels is interrupted in droughts, just when it is most required to observe minimum water table levels and the aquifer response to pumping.

Building an element of "drought resistance" into water- supply programs has included the provision of extra well-lining rings for future deepening, drilling a limited number of extra boreholes in favorable strategic locations to be uncapped and used in emergencies, and ensuring adequate borehole depth. In India, experience of drought impacts on groundwater sources led UNICEF to recommend a minimum drilling depth ("norm") of 60 m for hand-pump boreholes in the hard-rock areas to allow for large drawdowns in dry conditions.

## Intrinsic Groundwater Quality Problems

### *Hydrogeochemistry and Health*

Nine major chemical constituents (Na, Ca, Mg, K, $HCO_3$, Cl, $SO_4$, $NO_3$ and Si) make up 99 percent of the solute content of natural waters. These constituents provide the hydrochemical characterization of waters and their proportions reflect the geological origin (type of rock), groundwater flow paths and history of groundwater.

Elevated concentrations of solutes can occur in certain hydrogeological environments, such as increases in salinity due to evaporative concentration, high sulfate concentrations associated with weathering of basement rocks, dissolution of evaporites in sedimentary sequences, hardness associated with carbonate rocks, and from association with some types of geothermal activity. The key objective of groundwater quality monitoring programs is to zone areas where groundwater is unsuitable for potable supply. It should be noted that concentration of some of the above constituents can be increased as a result of polluting activities at the land surface, and it will also be important for management to differentiate human impacts from natural quality problems.

Reactions between rainwater and bedrock during percolation provide groundwater with its essential mineral composition (Freeze and Cherry, 1979). Rainfall infiltrating through the soil takes-up carbon dioxide and the resulting weak carbonic acid dissolves soluble minerals from the underlying rocks. In humid climates with significant recharge, groundwater moves continuously, contact time is short and only the most readily soluble minerals will be dissolved. Groundwater in outcrop recharge areas in such regions is likely to be low in overall mineralization compared to that in arid or semi-arid regions in which the combination of evaporative concentration and slow movement can produce much higher concentrations. Groundwater in igneous rocks, for example, is often of exceptionally low mineralization because groundwater movement is via joints and fractures and many such rocks are highly insoluble.

Minor and trace constituents make up the remaining 1 percent of the total, and can sometimes give rise to health problems or unacceptability for human and/or animal use/consumption (Freeze and Cherry, 1979). Many trace elements are essential for human health in small quantities (Figure 9), and are taken in from both drinking water and food. The desirable concentration range is, however, small and some are harmful at slightly higher concentrations. Others are not essential for health but are also harmful at low concentrations.

Low concentrations of essential elements in drinking water can cause community health problems, particularly if supplements are not provided by a healthy diet. Perhaps the most important problem associated with drinking water are linked to *iodine deficiency*. It has been estimated that up to 1000 million people are at risk globally from iodine-deficiency disorders, of whom some 200-300 million are goitre sufferers and some 6 million are affected by cretinism. The rocks of the earth''s crust are relatively depleted in iodine, whereas the highest concentrations are found in the oceans. Maritime rain has adequate amounts of iodine compared to continental rain and the problem is largely one of the continental interiors.

*Fluoride* is also an element that is sometimes deficient, but in the provision of rural water supplies from groundwater, excess is more likely to be a problem. The range of desirable concentrations of fluoride in drinking water is relatively small. At concentrations below about 0.5 mg/ℓ, dental caries may result, and fluoride is added to many toothpastes and some water supplies to promote dental health. Concentrations above 2.0 mg/ℓ in drinking water can begin to cause dental fluorosis and above 5.0 mg/ℓ can cause crippling skeletal fluorosis. High fluoride concentrations in groundwater are quite extensive. In India, 20-60 million people are affected, as are those living in some hydrogeological environments of China, East Africa and the Middle East. High fluoride in groundwater is thus a fairly widespread, and usually underrated, constraint on the provision of rural water supplies.

The trace element of greatest concern, however, is *arsenic*, which is both toxic and carcinogenic. Toxicity depends on the form of arsenic ingested, notably the oxidation state and whether organic or inorganic. Arsenic intake may be larger from food, but drinking water represents the greater hazard since the arsenic is present in the inorganic form. The WHO guideline has recently been reduced from 50 to 10 µg/ℓ. Most drinking waters have arsenic concentrations well below this, but concentrations in excess of 1.0 mg/ℓ are recorded in some areas. Documented cases of chronic arsenic poisoning are known for a number of different hydrogeological environments in Taiwan, Chile, Argentina, Mexico, China, India and Bangladesh. The latter, which appears to be the most widespread, is detailed in Box 6.

There are also major quality issues linked to *soluble iron* and *nitrates* in groundwater. In many places, especially but not only in crystalline basement areas, high iron concentrations cause water supply acceptability problems. Under reducing conditions, concentrations of dissolved iron may reach several mg/ℓ, and the solubility is greater at the low pH values which prevail in such regions. As the water is drawn to the surface and encounters oxygen, the dissolved iron is oxidized.

There is also evidence (Lewis and Chilton, 1984; Langenegger, 1994) that the use of galvanized iron pump components or mild steel borehole casing can make the situation worse, by adding iron and zinc to the abstracted water. Often beneficiaries do not fully utilize affected supplies and go back to traditional sources which have low iron concentrations but very poor bacteriological quality. The occurrence of elevated nitrate (and sometimes ammonium or nitrite) concentrations in groundwater supplies is normally related to pollution from agricultural practices and/or sanitation arrangements and as such is dealt with in Chapter 4. However, elevated concentrations may occasionally occur in arid regions as a result of natural soil-plant processes.

### *Effects on Irrigated Crops*

The quality of irrigation water quality is judged against criteria based on the adverse effects of the constituents of the water on the growth and development of irrigated crops, on the soils which are being irrigated, on agricultural workers and on consumers of the harvested products. These criteria have been developed over time from data which have often been empirical rather than scientific. Field experience has shown that knowledge of the quality of groundwater to be used for irrigation is essential. The main water quality problems related to irrigated agriculture are salinity, constituents that reduce soil infiltration rate, the toxicity of specific solutes and miscellaneous effects such as excessive nutrients (Table 7).

**Figure 9: Major and trace elements in groundwater and their health significance**

| TRACE ELEMENTS | | | | MAJOR ELEMENTS | | |
|---|---|---|---|---|---|---|
| measurement requires expensive equipment | | | | mainly simple and cheap to measure | | |
| 0·0001 - 0·001 mg/l | 0·001 - 0·01 mg/l | 0·01 - 0·1 mg/l | 0·1 - 1·0 mg/l | 1·0 - 10 mg/l | 10 - 100 mg/l | >100 mg/l |
| Rb | Li | P | Sr | Mg* | Na* | $HCO_3$ |
| La | Ba | B | F* | K* | Ca | |
| V | Cu | Br | | Si | $SO_4$* | |
| Se* | Mn* | Fe* | | | Cl | |
| As* | U | Zn | | | $NO_3$* | |
| Cd* | I | | | | | |
| Co | | | | | | |
| Ni* | | | | | | |
| Cr* | | | | | | |
| Pb* | | | | | | |
| Al* | | | | | | |
| Y | | | | | | |

**ESSENTIAL ELEMENTS**

| | |
|---|---|
| Cu (solid box) | considered essential for human/animal health |
| Sr (dashed box) | probably essential for health |
| B | non-essential elements |
| * | also considered to be toxic or undesirable in excessive amounts |

**N.B.** 0.001 mg/l (or ppm) ≈ 1.0 µg/l (or ppb)

*Note:* The concentration ranges indicated are the normal levels of occurrence, but much higher concentrations may be encountered under certain conditions.
*Source:* Edmunds and Smedley, 1995.

*Salinity* begins to become a problem if salt accumulates in the crop root zone to a concentration which prevents the crop from extracting sufficient water because of the osmotic pressure, and the plant growth rate is reduced. Because of evaporation from fields and remobilisation of soil salts, soil water is usually 2-3 times (and often 5-10 times) more concentrated than applied irrigation water. The recommended restrictions on the salinity of irrigation water (Table 7) refer to typical grain and fodder crops, and vary with the water application method and the soil type. For certain tree crops (such as citrus, groves and date palms) and various other crops the salinity limits are lower.

## *BOX 6 : Natural Contamination of Groundwater with Arsenic in Bangladesh*

• Use of groundwater in Bangladesh for rural domestic water-supply has increased greatly over the last 20 years, and the shift from more traditional surface water sources has reduced the human health hazard related to pathogenic contamination. Nonetheless, the natural quality of groundwater cannot always be guaranteed and purity can be impaired through natural build-up of toxic trace elements (notably arsenic) derived from long-term reaction with minerals in host aquifers.

• In Bangladesh groundwater is abstracted from the hydrostratigraphically-complex alluvial and deltaic aquifer over various distinct depth ranges. The bulk of rural domestic water-supply is derived from the shallow aquifer (above 100m depth) of Quaternary age. Fine-grained alluvium covers much of the surface, acting in part as a semi-confining bed for the shallower aquifer restricting the ingress of atmospheric oxygen.

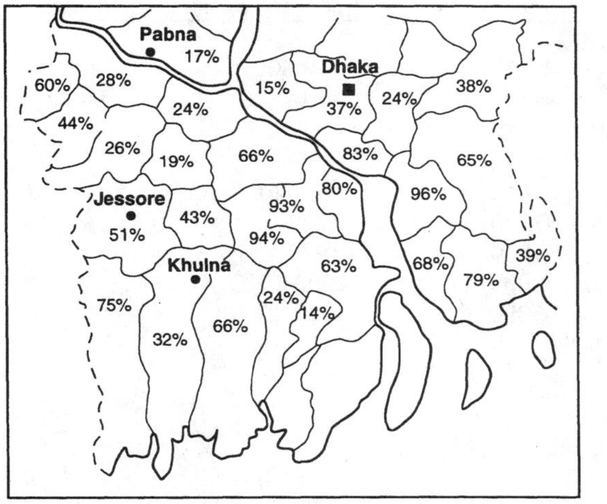


***Inset I:*** *Distribution of hazardous arsenic in shallow southern Bangladesh groundwaters (expressed as percentage of samples by district exceeding 0.05 mg/l in 1998 survey)*

• Groundwaters are almost entirely reducing and this is a key factor in the mobilisation of toxic concentrations of arsenic, in solution as both As(III) and As(V). A recent survey of more than 2,000 samples in southern Bangladesh (Inset I) has revealed that 35% exceeded 0.05 mg/l, while 50% exceeded 0.01 mg/l (the WHO recommended limit in drinking water) (BGS-MMD, 1998). However, groundwater from deeper Lower Pleistocene aquifers generally has lower arsenic concentrations (Inset II) and is being investigated as one alternative option for domestic water-supply.

• Chronic exposure to high concentrations of arsenic in water supplies gives rise to a number of severe health problems, including skin disorders (keratosis), as well as internal cancers, cardio-vascular and respiratory problems. The number of people in Bangladesh potentially exposed to drinking water exceeding 0.05 mg/l exceeds 20 million.

• The relatively high content of recent organic matter maintains the reducing condition of the aquifers, as a result of the limited supply of dissolved oxygen. This process also results in the reduction of most nitrate and sulphate, and high alkalinity following the generation of carbon dioxide. Bangladesh groundwaters also have relatively high concentrations of phosphorus and some exceed the WHO drinking water guideline for boron.

• The detailed mechanisms that give rise to the high arsenic groundwaters are not yet fully understood. Under anaerobic conditions, the reductive dissolution of iron oxides with release of bound arsenic is likely to be the dominant process. Lack of opportunity for flushing and oxidation of the shallow sediments in the current floodplain (as a result of their young age and the low hydraulic gradients) are also significant contributing factors. There are indications that both deep and shallow groundwaters in areas of geologically-older alluvial terraces have lower concentrations of soluble arsenic. Such areas are likely to have been subject to significant flushing by meteoric water during periods of low-stand of Quaternary sea level.

• Although the potential exposure to arsenic has increased in southern Bangladesh through increased use of groundwater, relatively few aquifers globally are impacted by such arsenic problems. It would therefore be unwise to abandon groundwater resources in developing countries in favour of alternatives such as surface water without proper hydrogeological and geochemical assessments.

| DETERMINAND (mg/l, except first two listed) | SHALLOW AQUIFER (<100m) | | DEEP AQUIFER (>150m) | |
|---|---|---|---|---|
| | median | maximum* | median | maximum* |
| pH | 6.7 | 7.0 | 6.5 | 6.9 |
| Eh (mv) | -3 | 77 | 32 | 98 |
| Ca | 6.5 | 74 | 5.0 | 18.3 |
| SO4 | <0.03 | 0.76 | 0.03 | 0.45 |
| HCO3 | 271 | 489 | 167 | 285 |
| NO3 | <1.3 | 4.4 | <1.3 | <1.3 |
| NH4 | <0.08 | 1.3 | <0.08 | 0.21 |
| P | 0.09 | 0.67 | 0.17 | 0.40 |
| B | 0.004 | 0.032 | 0.03 | 0.19 |
| Mn | 0.04 | 0.30 | 0.010 | 0.04 |
| Fe | 0.04 | 2.0 | 0.06 | 0.23 |
| As | <0.0005 | 0.030 | 0.0006 | 0.003 |

* maximum is represented by 5 percentile value (where values below analytical detection limit, half this limit has been used in statistical analysis)

***Inset II:*** *Groundwater chemistry of southern Bangladesh aquifers*

Leaching of salts below the soil zone is the key to controlling problems related to the applied water, but this requires the periodic application of excess low-salt water, which may not be feasible in groundwater irrigation. In practice, even if some leaching is possible, this may merely displace salt from the soil to underlying groundwater, and the concentrating effect of continuous recycling of salts is often observed in groundwater irrigation. In parts of northwestern Sri Lanka, intensive groundwater abstraction from a shallow sand aquifer has established a series of flow cells in which the abstracted water is applied to the ground and infiltrates back to the water table. Within this recycling system, groundwater chloride concentrations have risen by 100-200 mg/ℓ over the 20-30 year period of irrigation (Figure 4).

The two factors which affect the rate of infiltration are the salinity of the applied water and its sodium content relative to calcium and magnesium (SAR). High salinity water will tend to increase infiltration, but low salinity water tends to remove salts from the soil, reducing its stability. High sodium-to-calcium ratio promotes dispersion of soil aggregates close to the surface and smaller particles clog pores reducing water movement. The potential for these problems to occur is measured by the *electrical conductivity (EC)* and *sodium adsorption ratio (SAR)* of the applied water (Table 7).

**Table 7: Guidelines for interpretation of water quality for irrigation**

| *Potential irrigation problem*<br>*Harmful constituents* | *Degree of water use restriction required* | | |
|---|---|---|---|
| | *Low* | *Moderate* | *High* |
| SALINITY (affects crop water availability) | | | |
| Electrical Conductivity (EC-μS/cm) | <700 | 700-3000 | >3000 |
| Total Dissolved Solids (TDS-mg/ℓ) | <450 | 450-2000 | >2000 |
| INFILTRATION (soil infiltration rate) | | | |
| SAR = 0 - 3 and EC of | > 700 | 700-200 | < 200 |
| SAR = 3 - 6 and EC of | >1200 | 1200-300 | < 300 |
| SAR = 6 - 12 and EC of | >1900 | 1900-500 | < 500 |
| SAR = 12 - 20 and EC of | >2900 | 2900-1300 | <1300 |
| SAR = 20 - 40 and EC of | >5000 | 5000-2900 | <2900 |
| SPECIFIC ION TOXICITY<br>(affects sensitive crops) | | | |
| Sodium (Na-meq/l) | | | |
| - surface irrigation | < 3 | 3-9 | > 9 |
| - sprinkler irrigation | < 3 | > 3 | |
| Chloride (Cl-meq/l) | | | |
| - surface irrigation | < 4 | 4-10 | >10 |
| - sprinkler irrigation | < 3 | > 3 | |
| Boron (B-mg/l) | < 0.7 | 0.7-3.0 | > 3.0 |
| MISCELLANEOUS EFFECTS<br>(on susceptible crops) | | | |
| Nitrate ($NO_3$-N-mg/l)* | < 5 | 5-30 | >30 |
| Bicarbonate ($HCO_3$ -meq/l)<br>(overhead sprinkling only) | < 1.5 | 1.5-8.5 | > 8.5 |
| pH | <6.5 | 6.5-8.4 | > 8.4 |

* ammonia and organic nitrogen should be included where wastewater used

*Note:* This shows the restrictions recommended by the UN-FAO on water use to avoid damage to crops and/or soil for a wide range of chemical constituents

*Source:* Ayers and Westcot, 1985.

*Toxicity problems* occur if individual constituents from the soil or water are taken up by plants and accumulate to concentrations which cause crop damage or reduce yields. The degree of damage depends on root uptake and crop sensitivity. Perennial crops such as trees are more sensitive to toxicity problems. Ayers and Westcot (1985) provide considerable detail concerning crop sensitivity to toxic elements: the most important are Cl, Na and B (Table 8) but problems may also arise as a result of Se, As, Ba, Cd, Cr, Pb and Ni (mainly in situations where drainage waters are reused for irrigation or where wastewater reuse is involved, rather than irrigation from groundwater sources directly).

The last group of problems (Table 7) includes the effects of excessive nitrate content in the applied water, which may cause vegetative overgrowth, crop lodging and delayed maturity. High bicarbonate concentrations, which are not unusual in limestone aquifers, can cause unsightly deposits on fruit or leaves from overhead sprinkler irrigation, and low pH can encourage corrosion of distribution systems.

*Well Encrustation and Corrosion*

*Clogging* is an important cause of deterioration in borehole performance (Howsam, 1995). It is caused by the physical processes of redistribution of fine material in the aquifer, the gravel pack and infilling the borehole itself, and by chemical encrustation on screens and pumps. The deposits formed by both sets of processes reduce the permeability of the well screen, gravel pack and adjacent aquifer, and increase surface roughness, causing flow turbulence.

**Table 8: Key factors in the challenge of groundwater source maintenance for improved efficiency and useful life**

| | | *Relative importance** | |
|---|---|---|---|
| *Factors* | *Consequences/comments* | *Domestic wells* | *Irrigation wells* |
| TECHNICAL | | | |
| Quality of Design and Construction | • increases reliability of supply<br>• reduces need for major maintenance and rehabilitation | * | *** |
| Complexity of Wells and Pumps | • increases need for personnel training<br>• reduces opportunity for local maintenance and spares manufacture | * | **(*) |
| Accessibility of Area and Wellheads | • complicates logistics of energy supply, spares, etc.<br>• constraints on vehicles | ** | ** |
| HUMAN | | | |
| Ownership and Responsibility | • accountability needs to be clearly established<br>• community or user ownership preferred | * | ** |
| Operational Supervision and Organization | • ensure systematic monitoring and diagnosis<br>• procedures for supply of basic spares critical<br>• incentives for operational performance | * | **(*) |
| Personnel Training | • essential, especially for water users<br>• encourages user participation<br>• resolves cultural barriers | ** | *** |

*Note:* In practice many of these interact and overlap.

Clogging by chemical precipitation can affect the borehole, pump and pipework and the aquifer immediately around the borehole, reducing borehole capacity and pump efficiency. The most commonly-reported encrustations are those of iron oxyhydroxides, (sometimes in association with manganese deposits) and calcium carbonate. The former occurs (as described above) when anaerobic groundwater becomes oxygenated, causing conversion from ferrous to ferric iron and precipitation of insoluble ferric hydroxides. Precipitation of calcium carbonate is often quoted but less commonly observed. The contribution of microbial processes to clogging, either by enhancing iron reactions or by deposition of sludges or slimes of biological material (biofouling), is becoming recognized as important.

*Rehabilitation processes* for dealing with deterioration due to encrustation normally include the use of acids to dissolve calcium carbonate and iron hydroxides, including cases where microbial activity has contributed to the problem. The most effective agent for removing such deposits is hydrochloric acid, which is made up to 15 percent by volume and left in the borehole for up to 24 hours. For biofouling, biocides such as strong oxidizing agents and chlorine-based compounds are commonly used, but physical processes such as surging and jetting may be needed to dislodge biofilms from well screens so that the biocides can reach active bacteria (Howsam, 1995).

*Corrosion* of iron, steel and other metals (such as zinc) in aqueous solutions is essentially an electrolytic process involving anodic and cathodic areas in corrosion microcells. Large numbers of such microcells are present on metal surfaces due to differences in surface stresses, surface deposits, metal inclusions and other non-uniformities. Larger corrosion cells can result from differences in water temperature, flow conditions, concentrations of solutes, and also where dissimilar metals join to form galvanic couples. Electrochemical corrosion affects well casing and screens, pumps and pipework.

The quality of groundwater includes consideration of its pH (acidity) and Eh (oxidation-reduction) status. Many areas in which groundwater is drawn for rural supplies from the crystalline basement are subject to low pH and low concentrations of dissolved minerals. Groundwater is generally soft with high concentrations of free carbon dioxide. Such waters are indicative of the widespread corrosion problems described by Langenegger (1994) from several West African countries. In a comprehensive study of handpump components, 70 percent of groundwaters in the region were found to be corrosive. The effect is to impair mechanical performance by weakening pump rods and damaging rising mains. The cost-effectiveness of rural water supply schemes can be significantly affected by hand-pump corrosion. Not only will recurrent costs be increased by the necessity for frequent repairs, but the investment in boreholes and pumps may be wasted if the quality of the water produced renders them unused. Langenegger (1994) suggested that handpumps drawing water with over 5 mg/ℓ soluble iron are generally not used for drinking water, and 60 percent of boreholes in the West African study area had iron concentrations which at times exceeded 10 mg/ℓ.

Corrosion and encrustation processes are complex and interactive and, for this reason, no single test or index is an infallible indicator for predicting borehole life. However, because corrosion of screens and pumps is such an important cause of deterioration, some effort has gone into the development and use of indices and tests to assist in selecting construction materials and predicting borehole performance. The main direct measurements used are pH, free $CO_2$ saturation and stability indices, chloride and sulfate ratios, together with corrosion-resistance meters. The key to overcoming corrosion problems lies in the choice of construction materials (plastic, fiberglass or stainless steel) for screens, casing and pump components. In the rural water supply sector, much effort has gone into the development of corrosion-resistant rising mains and rods, since where groundwater is moderately-to-highly aggressive ($pH < 6.5$) galvanizing of mild steel does not protect from corrosion.

**Table 9: Analysis of factors reducing well efficiency and useful life**

| MANIFESTATION OF PROBLEM | TIMING OF CAUSES AND/OR OCCURRENCE | POTENTIAL UNDERLYING CAUSES | | | CORRECTIVE ACTION(S) | RELATIVE IMPORTANCE | |
|---|---|---|---|---|---|---|---|
| | | *Description of process* | *Physical* | *Chemical* | | *Domestic wells* | *Irrigation wells* |
| Excessive Pumping Costs | commissioning | casing-off productive aquifer horizons | • | | DC | • | •• |
| | commissioning | incomplete well development | • | | DC/OM | • | •• |
| | progressive | well screen/pump encrustation | O | • | OM/DR | • | •• |
| | progressive | aquifer overexploitation | • | O | AM | | • |
| Premature Pump Failure | commissioning | non-vertical borehole | • | | DC | • | • |
| | commissioning | inappropriate pump specification | • | O | DC | • | •• |
| | commissioning | sand pumping due to inadequate well-screen/gravel pack | • | O | DC | | • |
| | | pump pitting due to air entrapment during excessive drawdown | (see above for causes) | | | | • |
| | progressive | excessive wear of pump bearings | • | | OM | • | •• |
| | progressive | pump corrosion | | • | DC | • | • |
| | progressive | sand pumping due to well screen corrosion | O | • | DC | | •• |
| Well Siltation & Collapse | commissioning | inadequate well screen/gravel pack | • | | DC | •• | •• |
| | progressive | well screen corrosion | O | • | DC | • | • |

*Note:* Primary (•) and secondary (O) causes are distinguished where appropriate.
DC improve future well design and/or construction.
OM improve routine well operation and maintenance practices.
DR employ well diagnosis and rehabilitation techniques.
AM improve aquifer management/groundwater abstraction controls.

## Operational Sustainability of Rural Water supplies

*Maintenance—Functional Sustainability*

All waterwells need to be properly operated and carefully maintained if they are to sustain their yields and efficiency. Arlosoroff and others (1987) emphasized this, saying that "success or failure depends primarily on one factor, whether the new water supply can be maintained." Yet the proper operation and adequate maintenance of groundwater supplies is frequently overlooked until something goes wrong. When this happens, it may be too late to retrieve the situation effectively. It is always likely to be more expensive to resolve than to prevent a problem, and is also likely to take considerable time during which the water supply source may be inoperable. A major effort was made in the late 1980s to increase the attention given to operational sustainability issues, and efforts to bring about widespread improvements are still on-going with varying degrees of success.

Operational sustainability involves institutional, legal, financial, social and cultural issues, as well as a broad group of technical factors, and the interaction between them. The first major challenge is to meet the cash-flow requirements of operation and maintenance. There are legal and administrative issues to be resolved, including:

- Who owns the source?
- Who is ultimately responsible for ensuring the source is operational?
- How can they be held accountable?
- Who collects the fees and how is financial accounting done?
- What is done with the fees collected?
- Who is authorized to order spare parts and/or to install them?
- How can their availability be ensured?
- What external support is required or desirable?

The key issues are summarized in Table 8.

*Technical issues* also need to be addressed, in particular construction standards, maintenance equipment and personnel training. If boreholes are poorly constructed, they are more difficult and costly to maintain, will deteriorate in efficiency and may become contaminated. The importance of ensuring sound selection of well sites and tailoring design and construction to local hydrogeological conditions (including the use of screens and gravel packs if required) cannot be over-emphasized in the search for operational sustainability in rural areas.

Many domestic water supply schemes are relatively small in scale and often obtain groundwater from shallow aquifers, typically from depths of less than 100 m and often less than 30 m. While the total investment in large numbers of small rural water supply sources is high, the impact of temporary (or even permanent) loss of some individual sources, while causing increased local hardship, may represent relatively little overall loss of investment.

Irrigation schemes, on the other hand, utilize high-yielding wells from both shallow and deep aquifers, reaching several hundred meters. These require much more sophisticated design, construction, operation and maintenance, and have much higher capital and operating costs. They are more susceptible to externalities (such as power fluctuations or interruptions), government subsidy policies (affecting choice of fuel), and impacts of drought and changes in surface water availability for conjunctive use.

Maintenance of boreholes and pumps for irrigation is not substantially different from that for domestic water supply. Some factors, however, have greater relative importance (Table 8) and there are additional factors which may have an impact. In some irrigated regions, boreholes may be widely scattered among flooded fields with limited access. The availability of energy for pumps (electric or fuel), the accessibility for regular inspection and maintenance may be limited and the degree of supervision may be less than for domestic supplies, where the consequences of supply problems may be instantly discernible and quite rapidly reported. In irrigation systems, individual farmers may suffer severely from a failing borehole but may not receive the rapid response to their needs that domestic consumers can often obtain. Unavailability of water at crucial times in the growing season may mean total or partial crop loss.

More sophisticated organization is required to operate and maintain large irrigation wells properly and more competent water-user associations are required if government wishes to hand over responsibility. In some countries there are a complex mix of public and private irrigation boreholes and this may further complicate the situation. The scattered rural nature of the installations and often poor road access increase the difficulties of monitoring inspection.

A properly operated and maintained waterwell should have a long *working life* or be essentially permanent. In Iran many ancient "*qanats*" have been operating for centuries, because they were well constructed and carefully maintained. In general, a new groundwater source should be sustainable if the "recommended yield" is not exceeded and if equipment is properly maintained and replaced as necessary. This care and attention is not always provided and wells fail through over-pumping, pump encrustation, corrosion or contamination.

The development of deep irrigation boreholes has widely facilitated cheaper food production and thereby contributed to poverty alleviation. However, in certain hydrogeological environments (notably thick multi-aquifer alluvial systems), they have led to permanent or seasonal lowering of the water table in the phreatic aquifer and caused failure of shallow domestic wells, reintroducing inequity of access to drinking water (World Bank, 1998). In such situations, effective resource regulation would require compensation (in terms of money or water) to those so affected.

Maintenance costs have important *hydrogeology-related components*, and correct diagnosis of these is critical (Table 9). Poor construction leads to hydraulically- inefficient boreholes and this produces excessive drawdowns, high pumping lifts and heavy wear on pump components. Sand pumping due to poor design or installation of screen and/or gravel pack also leads to heavy wear on pump components and infilling of boreholes. The former increases pump maintenance costs and the latter may produce total borehole failure, as in the case of the Lower Indus Valley of Pakistan (Box 5).

*Borehole construction standards* are of particular importance, and there is generally inadequate investment of public funds in this regard. It is essential that waterwells are properly constructed, that screens and pumps are correctly installed and that the wellhead is properly sealed to protect against direct ingress of polluted water. Without these essential elements, groundwater supplies are unlikely to be long-lasting. In the national rural water supply program in India, for example, simple and economical borehole designs which are generally suited technically to the local hydrogeological conditions have evolved over time. However, the understandable drive to achieve full national coverage of the rural population places such pressure on state implementing agencies that short cuts are often taken. The enormous task of field supervision of some 800 government drilling rigs means that inspection to help prevent poor standards of construction is inadequate.

This situation is compounded by the increasing use of private drilling contractors without adequate supervision to help meet coverage targets; they now construct about 80 percent of all new rural water supply boreholes. Thus a significant proportion of the 3 million or so domestic boreholes and 6-10 million irrigation boreholes will have been poorly constructed, perhaps in several respects. The outcome will be shortened life and/or increased operating and maintenance costs. However, without adequate monitoring programs neither the degree to which design lives (20 years) have been shortened nor the proportion of boreholes affected is known. Thus, the cumulative national cost of poor construction standards can neither be assessed or addressed.

*Preventive maintenance* of waterwells is not yet a part of the local culture of many countries. Staff who undertake maintenance and repair of pumps may not adequately maintain the vehicles they drive and coming to terms with the maintenance requirements of modern machines is a process which is going to take a few more years, especially in remote areas. Future projects need to target support at this area, and make training and education a more integral part of project implementation.

### *Cost Recovery—Financial Sustainability*

***Village Water Supply.*** Various ways of financing new groundwater sources may be available, including loans from commercial banks, central or local government support, and international grants/loans. Water user charges, however, are a critical factor, not only for cost recovery and helping to ensure sustainable operation, but also to obtain committed involvement and participation of the users. Water has so often been available at heavily subsidized prices, or essentially as a free good, that the move to raise prices towards economic levels is widely resisted and politically very difficult in some countries. There are still many systems and supplies worldwide that do not even recover a significant proportion of the *costs of operation, maintenance and distribution.*

Approaches have been tried to establish "fair prices," "politically acceptable prices," or "economically satisfactory prices" (Dinar and Subramanian, 1997). Surveys of "willingness to pay" are frequently used to help justify economic charges, but the fact remains that in many developing countries the poor pay the most in real terms for water and they often receive the poorest service. Their payment may be in the form of labor as well as money. Water is an essential of life, and people will pay whatever they are able to get it. Since the poor in general have to pay proportionally the most, the importance of increasing access of the poor to potable groundwater supplies cannot be over-emphasized as a major contribution to the alleviation of poverty.

It is now broadly accepted that the most effective means of achieving sustainability is to involve communities and to have them assess and collect water charges in sufficient amounts to at least cover the costs of operation and maintenance. Capital costs have hitherto often been fully provided by governments or aid agencies, but it is now becoming normal for projects to at least investigate the potential for communities to contribute to the capital cost and, if so, to collect an "up-front contribution" so that community commitment is ensured. The full or partial costs of maintenance are now often covered by the recipient communities, and the costs of spares procurement and distribution need to be built in. A further level of commitment is required to incorporate sufficient accumulated balances to cover predictable replacement of pumping equipment every decade or so. Various methods of collecting charges have been tried (Dinar and Subramanian, 1997) but it is important that the community develop or adopt the one that they are most comfortable with and can implement most successfully.

In the mid 1980s, Arlosoroff and others (1987) wrote: "capital costs of various levels of service depend very much on local conditions but the relative costs of the different groundwater-based technologies are apparent even though the range of costs may be quite wide." With some 1800 million rural people in need of improved water supply by the end of the century, the extra costs of a high-level service can be justified only when beneficiaries are willing and able to pay the extra costs in full. Nevertheless, higher service levels have been provided even when communities or beneficiaries only contribute a fraction of the cost and such arrangements are not likely to be sustainable.

Experience with recovering operation and maintenance costs has been mixed. In the developing countries, irrigation operation and maintenance cost recovery ranges from a low of 20-30 percent by the India and Pakistan governments to a high of about 75 percent in Madagascar, where water-user groups responsible for collecting water charges and maintaining physical facilities have been established.

Collection, accounting and auditing costs also have to be covered. Collection in a small community is no small task if left to an individual, but where it has been established as a community responsibility, water charges can be collected more promptly than in urban centers. Considerable experience in community water supply development has been acquired in the last 20 years and techniques for setting-up and maintaining effective community-run systems are now well established (Subramanian and others, 1997). There is still much to be done, however, to ensure that these are widely implemented and effectively monitored. Most of these community supplies use groundwater and the recovery of costs for groundwater differs from surface water in so far as the actual abstraction costs may be higher and the wells may require more regular maintenance than surface water intakes. However, treatment, storage and distribution costs are likely to be lower.

***Irrigation Water Supply.*** There is still much variance in the prices set for different uses (Dinar and Subramanian,1997). Irrigation water is typically much cheaper than water for domestic supply, and there is often little relationship between the prices set and the actual availability of water. Subsidies, direct, indirect and geographical are common.

Countries have different reasons for charging for water; some wish to recover costs, some want to transfer income between sectors through cross-subsidy and others want to use charges to improve water allocation as a component of an overall water conservation and resource management strategy. Several countries are exploring unique pricing-related issues. Israel and Jordan are considering scaling prices for irrigation water of different quality (saline, reclaimed wastewater, fresh water), adjusting prices to reflect water supply reliability and implementing a resource-depletion charge. Several countries are considering adjusting charges to reflect regional differences in water supply costs.

There is an argument that prices for irrigation water be set to reflect opportunity costs. However, a more realistic immediate objective is to recover sufficient revenue to ensure the viability of water entities. Evidence from the field suggests that farmers are willing to pay for reliable supplies of water, but the practical problems of pricing for irrigation services are complex. Fees are often set on the basis of irrigated area, which is by no means a direct or reliable measurement of the water received. The water drawn from boreholes can be directly measured, and one way to circumvent some of the problems of area-based charging is to measure the water delivered to a water-user association which has been delegated responsibility for allocating amongst farmers.

The record of non-payment of fees for water reflects two problems: lack of political and managerial commitment and weak incentives to collect and limited willingness to pay because services are poor. Failure to recover costs and to reinvest in systems leads to a vicious circle in which service declines and consumers, in turn, become less willing to pay for the poor-quality service provided. Conversely, higher collection rates often reflect decentralized management and enforced financial autonomy, which in turn deliver a high-quality service for which users are willing to pay.

Concerted efforts are being made to move towards *full capital-cost recovery* so as to establish and sustain viable water entities, but there is considerable variation in the progress that has been made (Dinar and Subramanian, 1997). One experience quoted in this paper comes from Korea, were farmland improvement associations (water-users associations) are responsible for recovering costs from farmers for projects completed and transferred to them. They set irrigation charges at levels to cover all operating and maintenance costs and a share of the capital costs amortized over 35 years, with government providing a grant for whatever capital costs are not paid for by the farmers. This has been very successful in collecting charges from farmers and repaying the government loan. Experience elsewhere has shown that water-user associations may be slow to get established, but that once the benefits are apparent and the water is available, effective participation and substantially-improved cost recovery become possible.

### *Community Action—Social Sustainability*

***Community-Driven Objectives.*** There is now broad consensus among the developmental sector that rural water supplies can only be sustainable with the *full involvement of local people*. This raises the question of how best communities can determine their own objectives and achieve these within the range of available physical, financial and technical options. Groundwater development presents particular problems because of the often invisible and somewhat mysterious nature of the resource. Some communities have good understanding of groundwater potential but for others groundwater may not even be recognized as an option, or there may be little understanding of the special techniques required to operate groundwater supplies on a sustainable basis.

The traditional donor approach of sending professionals to developing countries to initiate, design and implement projects resulted in successes, but many projects have not been sustainable because the beneficiaries were not fully involved so as to ensure continuation after disbursement ceased. From the 1980s, as expertise gradually became available in-country, national professional staff gradually supplemented and then replaced those from overseas. Even if these new professionals were potentially more sensitive to local capabilities, both the national institutional framework and the donors approach meant that local people were still rarely involved and little attention was paid to local experience at community level. It is now clear that it is the *role of the community* itself to determine what its priorities are. Outsiders can provide technical or other support and outline alternatives for them, but the decision has to be with the community.

Having decided what they want to do, the next task for the community is to realize it. There has been substantial funding available in the past from donors and from NGOs to support the construction and other capital costs of establishing water supplies. It is now recognized that the beneficiaries should invest when (and so far as) possible in the capital costs of source construction and equipping. In this changed situation, outside support is best provided in the form of training in the process of making appropriate decisions, and for technical and administrative training within the community, as well as for drillers, plumbers, equipment suppliers and installers. Tariff barriers on the importation of waterwell construction materials may also need to be removed.

**Table 10: The "integrated approach" to community groundwater supply planning**

| *Process* | *Condition or component* |
|---|---|
| • Effective Community Involvement | • in design, implementation, maintenance, financing<br>• communities wishes reconciled with capacity and willingness to pay |
| • Provision for Full Recurrent and Capital Cost Recovery | • support of capital costs for poorer communities only |
| • Maximum Use of National Services and Supplies | • in respect of drilling contractors, pumps, spares<br>• appropriate quality control to improve reliability |
| • Appropriate Technological Level | • Compatible with human and financial resources available |
| • Institutional and Manpower Development | • Closely mapped to needs of planned water supply system<br>• Parallel health/sanitation education |

*Note:* This remains a blueprint for improving operational sustainability of rural groundwater supplies.
Source: Arlosoroff and others, 1987.

In many instances, the NGOs led the way in working at the "grass roots level" helping communities to develop the type of groundwater supplies which they most wanted, and as far as possible at the locations which they most preferred. Even though NGOs were sometimes regarded with suspicion by governments, they were able to demonstrate the effectiveness of genuine community development based on the wishes and efforts of the rural people themselves. However, the process takes time and money, so that the rate of coverage is necessarily slowed down and the per capita cost is higher.

***Role of Water-User Associations.*** The establishment of *water-user associations (WUAs)* has been encouraged (or even required by more recent projects) with promising results, particularly in respect of maintenance of rural water supplies and water allocation within irrigation projects. It has been shown, however, that obliging people to form water-user groups is often less effective than providing support for a process of helping the community determine what their objectives really are, within a range of available options. When the water becomes available for use, then people more readily see the need for a mechanism for sharing it fairly. Some irrigation projects have had little success in encouraging effective user groups until the benefits are clearly within reach. On the other hand, there is much less difficulty if the project objectives have been designed to meet the expressed needs of an established community or user group.

Water-user groups can be set up in a variety of ways, and it is important that their characteristics should be selected by the participants to suit local capacities and culture. Partnerships, co-operatives, stock companies and bulk-supply companies have all been established in different locations and contexts. The more sophisticated are normally reserved for larger community supplies and for irrigation systems. It is absolutely essential that they are kept to a level of simplicity and transparency that is appropriate for the users themselves.

Water-user associations can contribute to better performance of irrigation systems because of their advantages over a public agency on the one hand, and over uncoordinated activity by individual water users on the other (Subramanian and others, 1997). Nevertheless it is safe to say that one cannot expect WUAs to achieve sustainable levels of system performance by themselves. Along with the institutional structure of WUAs a combination of appropriate technology, supportive state agencies/policies and economic forces (including clear property rights and profitability of irrigation enterprises) is required to sustain the WUAs themselves, as well as for sustainable irrigation systems.

With regard to *water supply and sanitation associations (WASAs)* "there are no ready solutions or instant methods of promoting sustainable water and sanitation service delivery." There are situations and contexts where WASAs are appropriate, but there are also cases where the institutional costs of operating through WASAs could be extremely high. A water and sanitation project manager planning to decentralize service provision and production through WASAs is therefore best advised to adopt a flexible "doing and learning" approach, rather than following a fixed blueprint or rigid guidelines (Subramanian and others, 1997).

For the purposes of this discussion, it is worth repeating a table by Arlosoroff and others (1987) on the integrated approach to community groundwater supply planning (Table 10). The principal elements that must be taken into consideration are highlighted, and they closely mirror the contents of this section. More than 12 years have passed, but the spread of the WUAs in the 1990s illustrates that serious efforts are now being made to improve the operation of groundwater supplies, whether for domestic or agricultural purposes. Much, however, remains to be done in all of these areas so as to achieve operational sustainability on a long-term basis.

## Resume on Groundwater Supply Development

Since the first modern guidance manual on rural water supply (Arlosoroff and others, 1987) there have appeared various other major reference works on this subject (UNDP-PROWESS, 1990; WHO-IWSC, 1993; DFID-WELL, 1998). These deal in depth with most of the technical, financial and social aspects of the development and maintenance of rural water supplies (summarized above) in relation to any type of water source.

However, none of these guidance manuals enter into any detail on the investigation and evaluation of the hydrogeological factors that control the availability of groundwater supplies to wells and springs and that influence the natural intrinsic quality and the vulnerability to pollution of these supplies. Given the general complexity and dependence on local detail associated with successful well siting, design and protection, it is essential to engage a competent hydrogeologist in the early stages of planning groundwater-based rural water supply development programs. This, together with the effective databasing of waterwell records and hydrogeological data, are key roles for government action.

Governments also still have a role to play in supporting the development of community objectives for water supplies. Several are now actively encouraging the establishment of water-user groups for both domestic and irrigation water supply. Approaches vary somewhat from the "top-down targeted style" to a "more subtle and supportive line." The latter is likely to be more effective as it involves the people to a much larger extent in a process which they can help design and maintain. Experience has shown this to be the key to operational sustainability.

# 3

# *GROUNDWATER RESOURCE SUSTAINABILITY*

## Context and Challenge for Management

### *Hydrogeological Constraints on Resource Availability*

All groundwater abstraction by wells results in some decline in aquifer water level (or piezometric surface) over a certain area. Some reduction can be considered not only as necessary but also desirable since it often improves land drainage and maximizes groundwater recharge rates, by providing subsurface storage space for the infiltration associated with high rates of excess wet-season rainfall.

However, if the overall abstraction rate in a given area, or aquifer system as a whole, exceeds the long-term average rate of replenishment, there will be a continuous decline in water level, *overdraft or mining of aquifer storage* and consumption of aquifer reserves. The same applies to abstraction from deeper semi-confined aquifers in which the long-term rate of leakage induced to flow through the confining beds from overlying shallow aquifers is less than the abstraction.

An important factor which should constrain abstraction is the need to maintain groundwater levels in, and discharges to, the surface water environment (for example, groundwater-fed wetlands and brackish coastal lagoons), because of ecological, commercial and/or recreational interests.

For groundwater abstraction to be regarded as sustainable the constraints imposed by aquifer recharge rates must be respected, albeit that there may be significant difficulty in estimating these with adequate precision (Foster, 1992). There are a number of significant complications:

- General uncertainties about aquifer recharge mechanisms and rates as a result of inadequate field data
- The area for which the groundwater balance should be evaluated, especially in situations where pumping is very unevenly distributed
- The period for which this balance should be evaluated, especially in the more arid climates where major recharge episodes may occur as infrequently as once a decade or even once a century.

The way in which the latter two factors are interpreted in practice will vary considerably with the storage volume of the aquifer system, and its propensity to irreversible side effects as a result of short-term overdraft. Both will be a function of aquifer type and hydrogeological setting. Small (very localized) aquifer systems with low storativity and recharge rates will give rise to the most immediate concern.

### *Consequences of Uncontrolled and Excessive Abstraction*

*Groundwater resource (or aquifer) overexploitation* is an emotive, but useful, expression (Foster, 1992). Although not capable of precise scientific definition (for the reasons given above), groundwater scientists and water resource managers must realize that it has clear register at the political level.

However, in practice, we are more concerned about the consequences of abstraction than with its actual level. These include reversible interference with other wells and with springs, but can also include quasi-irreversible aquifer degradation due to ingress of saline or polluted water (Custodio, 1992; Foster, 1992; Llamas, 1992; Collin and Margat, 1993). In reality, there is a wide range of exploitation-related effects (Table 11) and it is not always appreciated that differing hydrogeological environments show varying susceptibility to the *side effects of excessive abstraction* (Table 12). Such side effects will, in many instances, be difficult to predict with precision until some systematic monitoring of aquifer response to abstraction has been undertaken.

**Table 11: Consequences of excessive groundwater abstraction**

| | *Consequences of excessive abstraction* | *Factors affecting susceptibility* |
|---|---|---|
| *Reversible Interference* | • pumping lifts/costs increase | - aquifer response characteristic |
| | • borehole yield reduction | - drawdown to productive horizon |
| | • springflow/baseflow reduction | - aquifer storage characteristic |
| | • phreatophytic vegetation stress (both natural and agricultural) | - depth to groundwater table |
| | • aquifer compaction/ | |
| | • transmissivity reduction | - aquifer compressibility |
| *Irreversible Deterioration* | • saline water intrusion | - proximity of saline/polluted water |
| | • ingress of polluted water (from perched aquifer or river) | |
| | • land subsidence and related impacts | - vertical compressibility of overlying/interbedded aquitards |

*Note:* The two effects in the middle band may be either reversible or irreversible depending on local conditions and the period during which the excessive groundwater abstraction persists; the immediate groundwater level response to abstraction and the longer-term trend will be controlled respectively by the aquifer response characteristic (ratio of transmissivity to storativity) and the aquifer storage characteristic (ratio of storativity to average annual recharge)
*Source:* Foster, 1992.

**Table 12: Susceptibility of hydrogeological environments to adverse side effects during excessive abstraction**

| | *Type of side effect* | | |
|---|---|---|---|
| *Hydrogeological environment* | *Saline Intrusion or Upconing* | *Land Subsidence* | *Induced Pollution* |
| *Major Alluvial Formations* | | | |
| * coastal | ** | ** | ** |
| * inland | * (few areas) | * | ** |
| *Inter-Montane Basins* | | | |
| * with lake deposits | ** (some areas) | *** | * |
| * without lake deposits | * (few areas) | * (few areas) | * |
| *Consolidated Sedimentary Aquifers* | ** (some areas) | - | * (few areas) |
| *Recent Coastal Limestones* | *** | * (related solution features) | * |
| *Weathered Basement Complex* | - | - | * |

*Note:* The number of asterisks gives a relative indication of the severity of potential side effects, which may only be of restricted geographical distribution in some instances.

Significant *reversible side effects* (well pumping cost increase and yield reduction) occur if an excessive number of boreholes are drilled in relation to the available resource and its optimum exploitation pattern, and this can be particularly marked where the hydraulic structure of an aquifer is such that its most productive horizons occur at shallow depths and are thus prone to early dewatering (Figure 10). While such effects are essentially reversible in a physical sense, their consequences upon groundwater users may be terminal, bearing in mind the time scales involved.

More serious are *near irreversible side effects*, especially those involving the *encroachment of saline water* (UN-FAO; 1997). This may intrude laterally from the sea (Figure 11), if coastal hydraulic gradients are reversed, but rather commonly also occurs from above in layered coastal aquifers. Such aquifers often have a strong upward component of natural hydraulic gradient which may reverse with pumping from deeper freshwater horizons inducing the ingress of overlying saline water.

**Figure 10: Progressive deterioration in operational performance of a production borehole in a heavily abstracted alluvial aquifer**

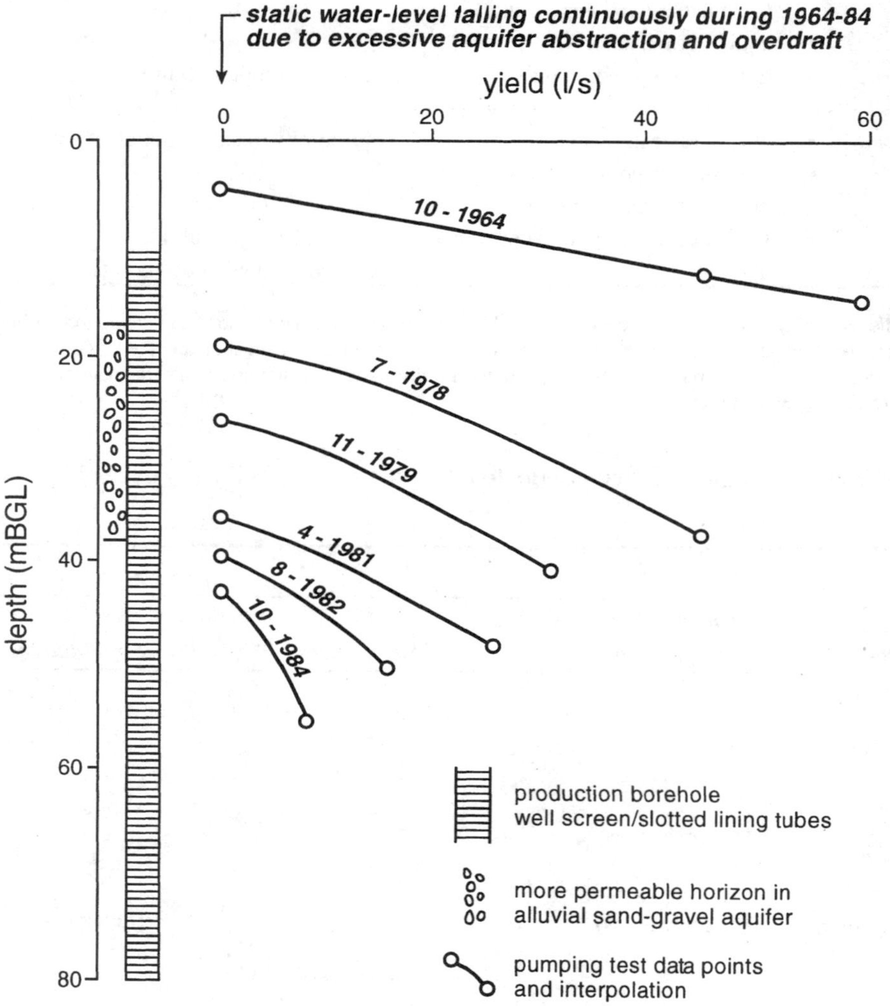


*Note:* As a result of dewatering of the most productive horizon due to excessive abstraction, maximum yield decreased from 60 to 10 l/s during 1964-84 while pumping lift increased from 15 to 55 m.

**Figure 11: Dewatering of groundwater storage in the Tertiary limestone of southeastern Cyprus due to intensive uncontrolled development for agricultural irrigation**

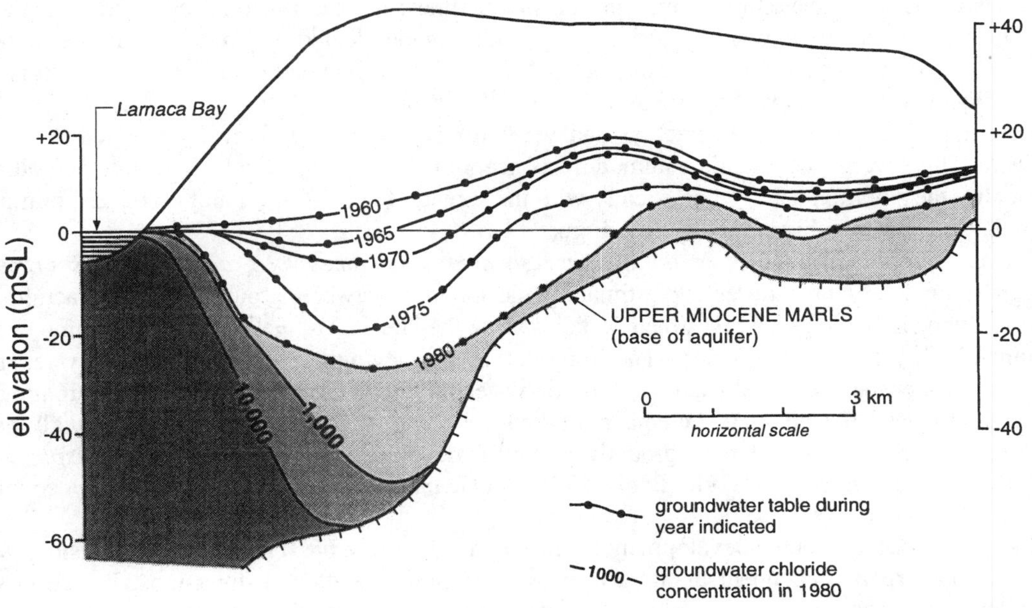


*Note:* Over a 20-year period major reduction in the saturated aquifer thickness has occurred as a result of both water table decline and saline intrusion.

The effects are quasi-irreversible since saline water, which first invades macropores and fissures, diffuses rapidly into the porous aquifer matrix under the prevailing high salinity gradients (Foster, 1992). It will then take decades to be flushed out, even after the flow of fresh water has been re-established. The ingress of saline water is terminal for virtually all uses, and can also result in damage to overlying soils if farmers continue to irrigate with increasingly brackish water in an attempt to obtain some return on their investment in wells.

By way of contrast, inland thick alluvial and sedimentary-basin aquifers in the more humid climates exhibit much less risk of significant exploitation-related side effects. However, even here there is the potential problem of increasing social inequity if deeper, larger-capacity irrigation boreholes lower the regional water table and reduce access to water supply for users of shallow domestic wells.

There are some who have argued that economic constraints, imposed by free market competition, are the only effective control over groundwater abstraction (Young, 1992). The larger capital cost of completing wells of increasing depth and decreasing yield, and escalating recurrent cost associated with pumping from ever greater depths will, it is suggested, rapidly result in achieving an optimum level of resource development and more efficient use of the groundwater produced.

If the only *externalities of groundwater exploitation* are hydraulic interference with other groundwater users, then this approach may be tolerable. Although even here there is failure to recognize the cost of drilling a disproportionately large number of wells to greater than optimum depth for the overall yield obtained, and the range of social and environmental costs associated with groundwater level declines (Foster, 1992). Moreover, social inequity may be further aggravated when the access to groundwater supply of those dependent upon shallow wells is compromised by water table lowering through heavy abstraction from deep, high-capacity, irrigation boreholes. The position is likely to be far worse if unrestricted abstraction causes quasi-irreversible aquifer degradation, most notably if increases in salinity are involved.

*Socioeconomic Problems and Obstacles*

***Historical and Political Perspectives.*** Groundwater abstraction is not new, but development on the large scale is. Wells have been excavated ever since pre-historical times, but the potential changed radically as advances in rotary drilling technology, in the turbine pump and in geological knowledge spread, most notably during the 1960s and 1970s. This is so despite the fact that churn drills originated in early Chinese history and percussion techniques were developed by the Flemish in the 12th century.

Early techniques of groundwater abstraction had very limited capacity and by comparison *resources appeared infinite*. This divergence led to a common misconception which lies at the heart of overexploitation concerns. In reality the situation changed drastically with the spread of deep drilling and motorized pumping, but perception lagged considerably behind reality (Figure 12).

In many areas, government policies encouraged *unrestricted development of groundwater resources*. In India, although rudimentary procedures for estimating the balance between groundwater abstraction and recharge have been available to guide investment policies since the late 1970s, well drilling and pump energy sources remained highly subsidized (Box 7). The irony is that a policy aimed at making groundwater more economical for all is a primary cause of shallow domestic water supply wells drying up, thus exacerbating the economical access to water for the poorest members of the rural community. Furthermore, virtually all Indian government organizations concerned with groundwater were developed to *promote resource exploitation rather than resource management* (World Bank, 1998). Such patterns were repeated in many countries worldwide.

Large-scale groundwater resource development for irrigated agriculture has a relatively short history when compared to its counterparts dependent upon surface water impoundments and diversions. Irrigation with surface water resources was one of the key historical elements in the promotion of civil society, because it generally needed cooperation amongst water users, and between water users and state governments, to make it possible. This was only locally the case for groundwater development, since it was often carried out on an individual (or small cooperative) basis and did not require the development of an effective institutional framework for water provision. Hence it now represents a considerable challenge to promote sustainable use of groundwater resources as a whole.

The challenges inherent in this history are compounded by the increasingly critical role groundwater resources play in the livelihoods of individual users in many developing nations. Access to groundwater for irrigation is making a very positive impact on subsistence and income for poor farmers, and in many cases also reduces the need for the rural poor to migrate during droughts by increasing income security (Chambers and Shah, 1989). These direct individual benefits make any subsequent constraints on groundwater use politically sensitive.

In some cases, governments have initially encouraged groundwater development for sound social and economic reasons to meet the needs of rural populations, albeit without consideration of the resource base (Reisner and Bates, 1990). This pattern is particularly well documented in the case of India, where electrical power for agricultural uses (which are dominated by groundwater pumping) is often supplied at nominal rates. In the longer-term this has imparted political legitimacy and popularity, and led to a situation in which policy reform initiatives have then been strongly opposed (World Bank, 1998).

## *BOX 7 : Critical Role & Future Uncertainty of Groundwater in Rural India*

• Groundwater is central to rural development and food security in India. Over 50% of the irrigated area is dependent on groundwater. Agricultural yields are generally 30-50% higher in groundwater irrigated areas. In addition some 85% of drinking water needs in rural areas are met from groundwater.

• These statistics, however, understate the critical role groundwater plays in the lives of rural inhabitants, since access to groundwater reduces agricultural risk and enables poor farmers to invest and to increase production (Chambers & Shah, 1989; World Bank 1998).

| STATE | 1968-69 | | 1984-85 | |
|---|---|---|---|---|
| | Private Dug Wells | Private/ Public* Tube Wells | Private Dug Wells | Private Public* Tube Wells |
| Andhra Pradesh | 0.660 | 0.016 | 0.982 | 0.085 |
| Bihar | 0.225 | 0.013 | 0.352 | 0.411 |
| Gujurat | 0.565 | 0.002 | 0.673 | 0.009 |
| Madhya Pradesh | 0.610 | 0.002 | 1.113 | 0.008 |
| Punjab | 0.170 | 0.114 | 0.091 | 0.595 |
| Tamil Nadn | 1.115 | 0.026 | 1.411 | 0.111 |
| Uttar Pradesh | 1.112 | 0.129 | 1.130 | 1.608 |
| **NATIONAL TOTAL** | **6.110** | **0.475** | **8.743** | **3.433** |

***Inset I:*** *Growth in waterwells (in millions) for selected Indian states (*public tubewells rarely exceed 5% of the category and are decreasing)*

• Development of groundwater resources in India proceeded rapidly (Inset I), and the number of energised wells has increased to more than 15 million in 1996 (Inset II). Increases in groundwater abstraction have had a major impact on the resource base in many arid and hard rock regions. Nationwide, the number of administrative 'groundwater resource blocks' classified as fully/excessively abstracted increased to 383 in 1992-93 (CGWB, 1995). The direct cost of groundwater overdraft in India to the end of the 1980s had been estimated at US$ 300 million, (almost certainly an underestimate).

• The emerging groundwater resource problems are closely related to high government subsidies in the agriculture sector. The subsidy on power-supply is perhaps the most significant where groundwater overdraft is concerned. In most states, power for agricultural pumping is provided at a flat annual rate based on pump capacity, and in some (such as Tamil Nadu and Punjab) is provided free of charge. Because of these high subsidies, power consumption in agriculture has grown dramatically. Official estimates indicate that it exceeds 40% of total energy use in many states.

• Addressing groundwater overpumping in India is complex. Centralised regulatory arrangements have existed since the 1970s, when a model regulatory bill was first circulated to state governments by the Central Groundwater Board. Although a few states have passed regulations, it is questionable whether they could be implemented given the millions of individual well owners on small land holdings (Dhawan, 1995), the inadequate administrative set-up and that reduction in subsidies generates strong political opposition.

• Nevertheless, some states are developing programmes, The Rajasthan State Government, (supported by the World Bank), is preparing a first phase initiative to build groundwater management capacity. This initiative combines investments in data collection with groundwater pilot projects to develop user-based management organisations in a series of groundwater resource conservation zones. It also contains major public education components.

• Where energy subsidies are concerned, Andhra Pradesh have taken bold steps to reform pricing structures, which should reduce the incentive for aquifer overabstraction. Other measures include:

* the prohibition of drilling deep tubewells for irrigation
* the mandatory construction of streambed groundwater recharge structures
* the introduction of economic incentives for dryland (as opposed to irrigated) cropping.

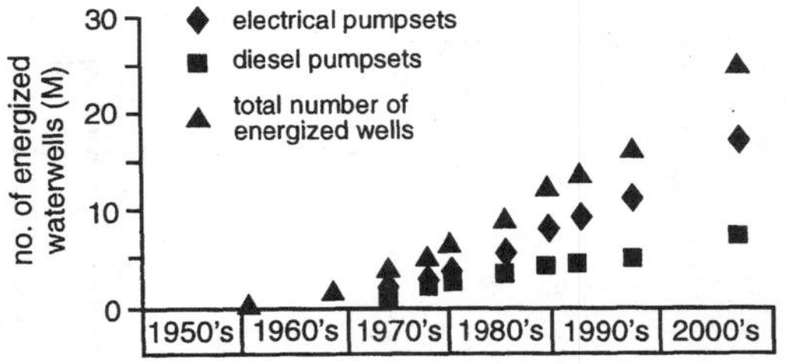


***Inset II:*** *Growth in number of energised waterwells in India*

**Figure 12: Historical development of the Deccan Traps groundwater system in Maharastra, India**

| YEAR | no. of wells (M) | average well abstraction ($m^3/d$) | est. total abstraction ($Mm^3/d$) |
|---|---|---|---|
| 1960 | 0.55 | 70 | 38 |
| 1970 | 0.70 | 100 | 70 |
| 1980 | 0.85 | 150 | 128 |
| 1990 | 1.25 | 230 | 276 |

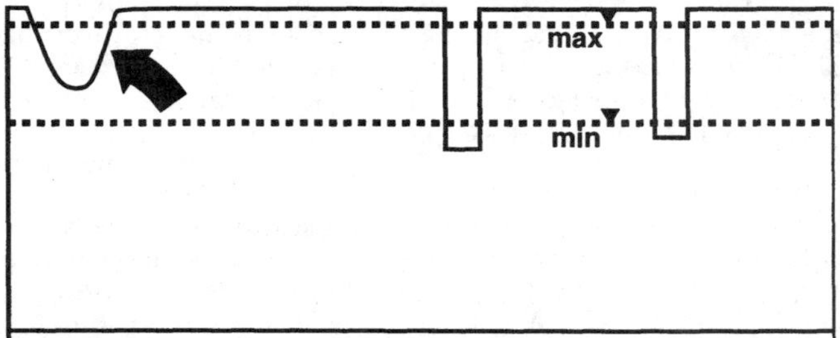


- limited groundwater abstraction (few wells/animal power/some diesel pumps)
- significant baseflow to streams

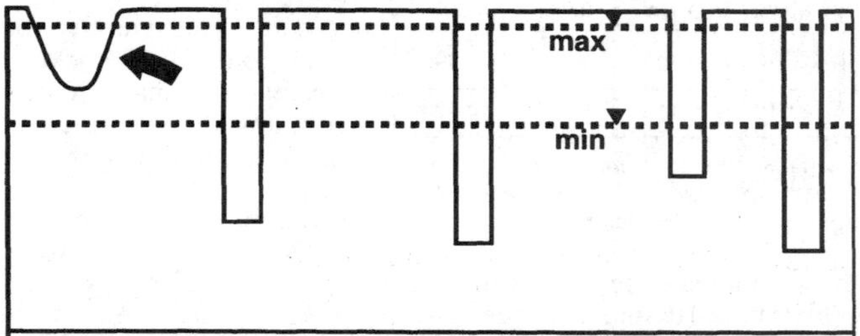


- increased groundwater abstraction (additional wells, diesel/electric pumps) and many wells deepened
- reduced baseflow to streams

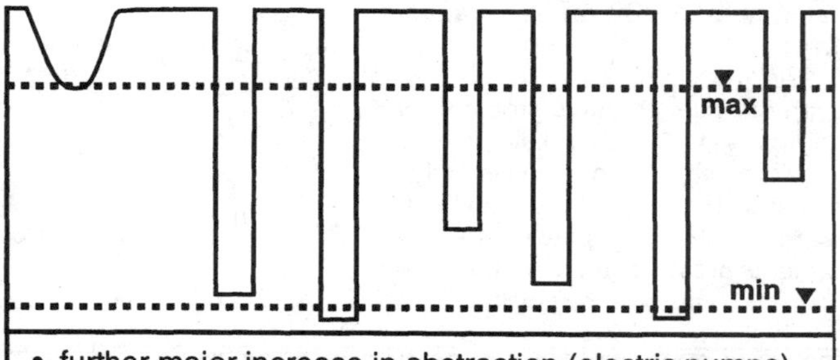


- further major increase in abstraction (electric pumps) with all wells deepened
- no baseflow discharge to streams

*Note:* The total abstraction increased 7-fold in a 30-year period as a result of both the spread of motorized pumping plant and of well drilling, but this led to intense competition for the available groundwater resources and virtual elimination of baseflow to local streams.

In sum, the historical perception of groundwater as an unlimited resource and the large benefits access to groundwater brings to individual irrigation users have been major factors underlying the rapid expansion in well numbers, the emergence of organizations focused on resource development, and the related political interests. In combination they represent both a cause of groundwater management problems and an obstacle to the development of effective responses.

***Economic Characteristics of Groundwater.*** The economic characteristics of groundwater resources have also played a major role in the emergence of management problems and represent significant obstacles to the development of management responses. Groundwater is generally undervalued (especially where its abstraction is uncontrolled) and there is a pressing need for national governments to undertake systematic valuation of their groundwater resources and for regulatory agencies to find ways of introducing economic instruments into resource management to begin to reverse this situation.

In this situation the user of the resource, in effect, receives all the benefits of groundwater development, but (at most) pays only part of the costs, usually the recurrent costs of pumping (although even energy supplies may be subsidized) and sometimes the capital cost of well construction (Figure 13). The economic costs associated with externalities (such as reductions in stream baseflows, impacts on wetlands, saline intrusion, loss of the strategic value of groundwater storage in extreme drought) are rarely included with charges to users, although they may suffer their consequences.

Moreover, in economic terms groundwaters (like fish) are a resource for which property rights are not readily and obviously defined in a legal sense (Foster, 1992). Thus, except in those nations where clear rights systems have been implemented, groundwater would still be termed a *common-property (or common -pool) resource*, which is to significant degree local in distribution. In this situation individual users have little ability to conserve groundwater for their own future use, and exploitation is notoriously difficult to control. It may even be subject to accelerated depletion, when individual users become aware of trends towards over-abstraction and attempt to recover their development investment while resources still remain.

The perception of groundwater as an infinite resource clearly contributes to undervaluation. Even where the finite nature of the resource is fully recognized, users often lack clear understanding of resource dynamics, aquifer boundaries and potential contamination, which further contributes to the *problem of undervaluation.*

Undervaluation is a key factor leading to economically-inefficient patterns of groundwater allocation and use (Young, 1992). In many cases groundwater is allocated to low value uses (such as the production of grain or fodder crops in arid regions), while higher value uses (such as provision of safe drinking water) are only partially met.

In addition, because the in-situ values associated with groundwater are rarely reflected in its cost, undervaluation creates a strong incentive for over-abstraction. This incentive is often further increased by direct subsidies and/or by indirect subsidies (such as crop price supports) that encourage allocation to lower value uses. At the same time, undervaluation reduces incentives for investment in water conservation and, more generally for resource management.

***Issues of Resource Scale and Variability.*** Rural groundwater use for irrigation is organized predominantly at the level of individuals or small groups, while aquifers range in scale from less than 10 $km^2$ up to major regional systems (that occupy areas of more than 1,000 $km^2$ and even 100,000 $km^2$).

Management tensions result from the fact that state and local administrative units, settlement patterns and cultural groupings rarely correspond to the *boundaries of aquifer systems*. In many cases there is a particular gap at the intermediate level. State organizations operate at a large-scale and find it difficult to address the highly localized factors governing groundwater use. At the same time, village and community groups lack the regional perspective and influence essential to understand aquifer management needs.

**Figure 13: Measuring the costs of groundwater abstraction**

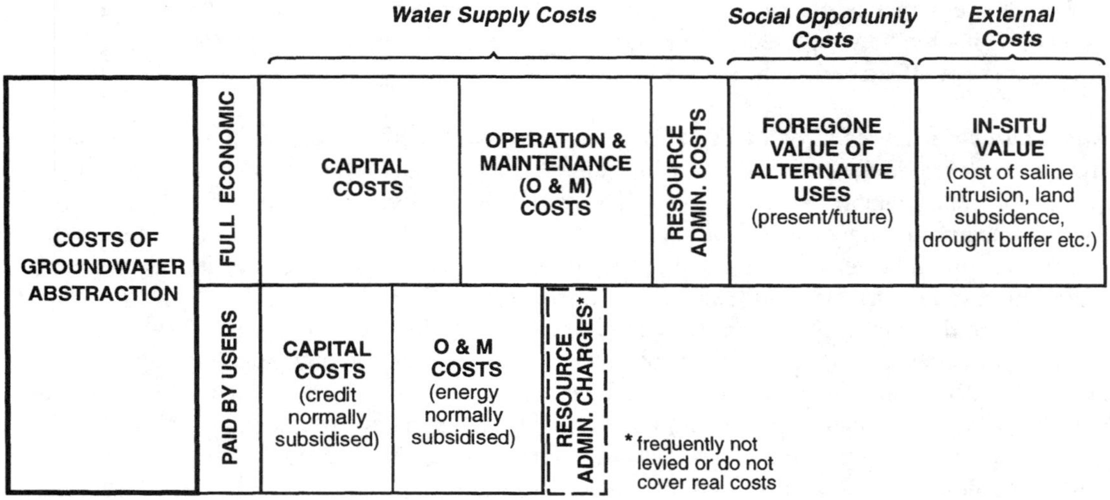


*Note:* No relative scale of full economic costs is implied, but it is evident that in the typical situation the costs paid by the user represent only a part of the full economic cost.

*Highly dispersed use patterns* can have significant aggregate impacts but the problems often arise at substantial distance from many users. The migration of a saline water front for example, is often due to changes in groundwater flow caused by regional pumping patterns, but the only users affected are those in the specific area where saline water intrudes.

In any one nation or region, there is likely to be significant variability in the hydrogeological factors controlling groundwater resource availability and the socioeconomic factors affecting their use for agricultural irrigation. These sets of factors interact and result in a *groundwater resource management context* that can change significantly over time and very greatly between locations. In India, for example, the national and state policies subsidizing both groundwater development and electrical energy for groundwater pumping have had little negative impact on the resource base in areas with adequate wet season rainfall and only seasonal groundwater use. But in the more arid areas these policies have greatly exacerbated aquifer overexploitation and lie at the root of unsustainable development patterns.

It is thus necessary to develop management approaches that can be tailored closely to specific situations which will vary with hydrogeological diversity. It also necessitates the development of adaptive management which can respond effectively as the larger economic or social contexts change. These needs conflict with the macro nature of many key policy tools. They also greatly reduce the chance of success for attempts to develop uniform management models or regulatory structures.

***Limitations of Regulatory Agencies.*** *National and regional regulatory agencies* are all too often under-resourced and weakly empowered when it comes to controlling groundwater abstraction. Moreover, all too often groundwater resource management remains under the administration of professional engineers who have been trained mainly in surface water resource development and tend to think in terms of major hydraulic structures rather than influencing large numbers of small stakeholders.

Simply increasing funding and empowerment, however, will not necessarily enable them to regulate groundwater abstraction effectively. A change of attitude and perspective is needed. There are also inherent factors which weaken their ability to introduce effective controls:

- The highly dispersed nature of groundwater abstraction, combined in many countries with deeply entrenched traditions giving individual landowners the right to abstract
- The uncertainty of most groundwater resource evaluation due to natural hydrogeological complexity and meteorological variability, and to inadequate monitoring of system response to abstraction
- The strong pressure for resource development (regardless of long-term consequences) sometimes exerted by the politically powerful lobby of land owners and/or plantation enterprises
- The high incidence of corruption of regulators during consideration of new abstractions or sanctions on illegal abstractions
- The lack of public and political awareness of the potentially irreversible consequences of excessive groundwater abstraction, and thus the absence of an adequate consensus for action.

The existence of *traditional methods of aquifer development*, often involving large numbers of small abstractors with shallow wells and limited pumping capacity in aquifer discharge areas, works in the contrary direction. *Implicit prior rights* held in perpetuity by such abstractors can make it difficult to introduce more rational use of groundwater and this tends to sterilize valuable storage resources against future development (Foster, 1992).

## Institutional Framework for Resource Management

To make effective progress groundwater resources management requires the effective integration of the key hydrogeologic and socioeconomic elements that determine and control the interaction between water/land-use and groundwater systems (Figure 14). For the purpose of groundwater management, the *institutional framework* is fundamental and consists of a set of organizations, social processes and legal agreements that enable management functions to occur. Perhaps the most important set of questions to ask in framing approaches to groundwater management are:

- What set of functions need to occur to address management needs within the specific context of concern?
- Are those functions already enabled adequately, either through formal institutions or through informal social processes?
- If not, how might the capacity of existing systems be strengthened, or new institutions be created, to enable these critical functions to be performed?

These questions can also be used as a starting point for framing *institutional reform programs*.

*Tiered Institutional Arrangements*

Groundwater management needs and options vary widely between locations, and thus have to reflect local hydrogeologic and socioeconomic conditions. However, local conditions often cannot be addressed in the absence of a higher-level enabling framework. In particular the need for adequate legal and/or social definition of groundwater abstraction rights (Feitelson and Haddad, 1998; Salman, 1999) will be a key provision in many instances.

Furthermore, many functions critical to groundwater management—such as determination of agricultural or energy subsidies—are a function of *national-level policy decisions*. As a result, institutional arrangements for groundwater management inherently involve multiple levels, and need to be tiered or "nested." Local management contexts are shaped by the dynamics of regional economic and institutional systems. Local institutions also draw legitimacy and authority for specific courses of action from a "higher level" or from *social norms*.

Key management functions are enabled by a chain of arrangements that connect basic principles (such as state ownership or trust responsibilities over groundwater) with the authority to act (for example by limiting

**Figure 14: General conceptual framework for the management and protection of groundwater resources**

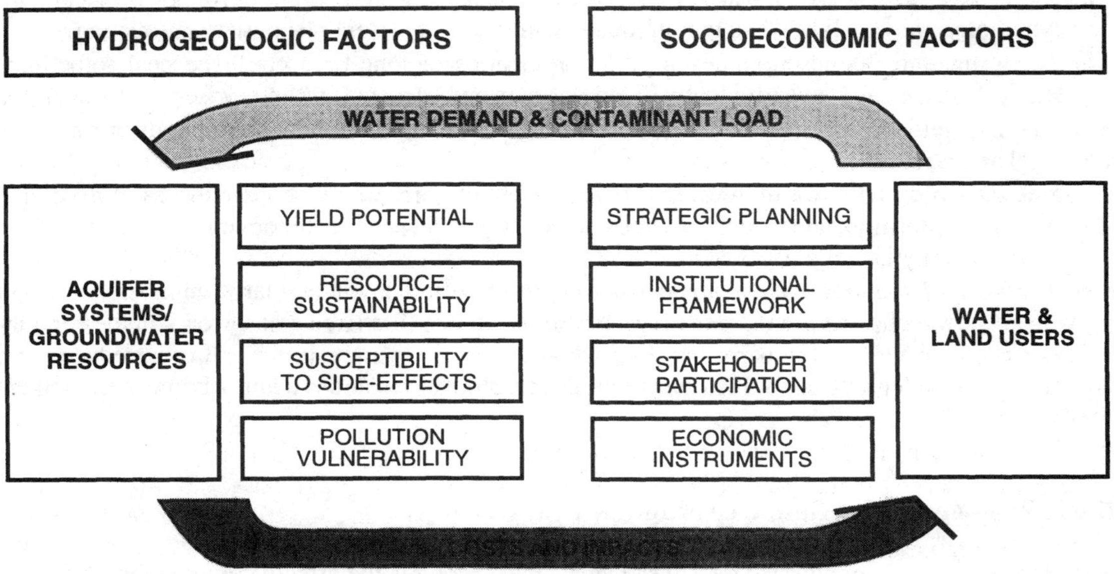


*Note:* The key hydrogeologic and socioeconomic elements that determine the level of impact and potential control of water/land use activities on groundwater are indicated schematically.

pumping from individual wells). These chains can evolve upward as management needs drive communities to develop higher-level enabling frameworks (such as legislation to create aquifer management committees) or they can evolve downwards from basic social principles to specific implementation policies (such as highly subsidized domestic water supply).

In general, the institutional framework shaping groundwater management options can be viewed as comprising four possible levels:

- The *macro high-level framework* comprising social norms, fundamental rights and legal principles
- State organizations, rights structures and market institutions operating at *regional level*
- *Intermediate level organizations* operating at the level of hydrological units (catchments or aquifers)
- *Local institutions* operating at the level of groups of users or communities.

Enabling groundwater management to occur requires institutional arrangements at several of these levels. As a result, it is essential to think through the relationship between each level and their importance in relation to the specific management functions needed. In doing this it is important to recognize the several issues which are discussed below.

First, agencies at different levels can have a range of functional roles depending on their capabilities. There is no inherent reason, for example, why regulations must be implemented and enforced by state agencies rather than local organizations. The appropriate level for regulation to occur depends on both the object of the regulation and the capacity of organizations at given levels.

Second, macro frameworks matter, providing the ultimate legal authority for action. *Constitutional provisions* specifying groundwater as owned by the state but subject to private appropriation or as the common property of all users, play a major role in shaping management options and public attitudes (Salman, 1999). In Yemen, for example, groundwater is treated as common property under one interpretation of the Islamic code and many view water markets or water charging as inherently unethical. Recently attention has focused on the *public trust concept* as a mechanism for balancing public and private interests in water resources (Koehler, 1995). This enables the state to initiate management while leaving room for private use rights and water market operations. Furthermore, because water is being held in trust for the people (rather than being either private or sovereign property), the concept enables public participation in management.

Third, in most situations there is a major gap at the intermediate institutional level. National water laws generally exist and in many cases articulate basic principles clearly (Salman, 1999). State regulatory agencies also often exist, though their capabilities vary greatly. There is, however, often a major gap between these organizations and the level of community groups and local users. This gap is of critical importance because most state organizations do not have the implementation capability essential to manage at aquifer scale. At the same time existing local organizations generally cover too little area to be effective in groundwater management and generally lack the necessary financial, technical and administrative resources. Intermediate institutions, such as *aquifer management committees*, will thus be needed in many instances.

The establishment and operation of aquifer management committees has a number of critical aspects:

- Defining a sound legal basis for their operation and relationship with the regulatory agency and local government
- Providing an element of financial support for meeting facilities, at least in their initial stages
- Promoting balanced representation of the groundwater user community, bearing in mind that even the agricultural user sector may be heterogeneous in terms of dependence on groundwater, cultivation regime and income status
- Developing and funding a technical information and communication system with the regulatory agency
- Ensuring that the macro level water resources framework does not distort the agenda of the committee and accidentally convert them into a policy lobby group on other water resource issues (such as subsidized surface water transfer schemes) or even agricultural issues (such as subsidies for a given crop).

*Flexible Management Schemes*

The context for groundwater management is dynamic. This implies that institutional frameworks must be able to adapt to change. The contrast between an *enabling adaptive framework* and more rigid structures for groundwater management can be illustrated by the different approaches currently followed in India and the western states of the USA (Moench, 1994).

In India, model legislation authorizing government control over groundwater has been in place since the mid-1970s. This bill (versions of which have recently been passed in several states) essentially creates a highly centralized government groundwater authority and provides it with a limited array of regulatory and enforcement powers. The top-down regulatory focus provides relatively little flexibility for adjusting to local conditions.

In contrast legislation enabling the formation of groundwater management districts in parts of the USA authorizes a wide variety of functions and places their implementation under the control of locally-elected boards of directors. These boards, however, rarely have sole authority, their scope of action being limited:

- On one side by private rights and the ability of individuals to enforce these through the courts
- On the other side by state and federal laws, and the powers these laws give to government agencies.

A system of checks and balances emerges in which the elected groundwater district boards often have broad authority but adapt courses of action to local conditions and changing circumstances. The effectiveness of groundwater management is (at best) partial, but the adaptive approach has facilitated substantial improvements (Kromm and White, 1990).

In whatever manner they are achieved, flexible and adaptive institutions are central to effective groundwater management. As a result, in evaluating existing institutional frameworks (or designing new ones), the degree to which they enable implementation to be adapted to local conditions and to evolve as circumstances change should be a major concern.

*Stakeholder Participation and Governance*

The importance of user participation is increasingly recognized for effective water management (World Bank, 1994). *User participation* is particularly important in the groundwater case due to the highly dispersed nature of resource use and the role individual decisions play in determining management outcomes. Substantial literature is available on water-user organizations and will not be reiterated here. The distinction between low levels of user participation and *roles for users in governance of institutions* is, however, important to emphasize in the context of groundwater resources.

In most cases, addressing groundwater over-abstraction requires demand-side management—changes of individual use for irrigation that reduce total abstraction. These changes need to occur in activities that take place daily and affect both livelihoods and lifestyles. Users must, as a result, play a paramount role in management and groundwater regulators need to work with them "*collaboratively on analysis, collaboratively in setting objectives, collaboratively in creating strategy* and *collaboratively in formulating project tactics*" (World Bank, 1994).

Collaboration must involve a dialogue between groundwater regulators and local stakeholders in which both parties have power to determine courses of action, and not one in which regulators encourage communities to participate by acquiescing to predetermined courses of action. This distinction is of fundamental importance, since in many cases a gulf exists between the approaches advocated by government authorities and the perceptions of local users. This gulf can become a continuing source of tension and needs to be bridged if it is not to undermine the effectiveness of management initiatives.

## Key Management Functions

Groundwater management is inherently complex. This complexity can, however, be greatly reduced by systematically identifying what needs to be done for effective management. Once this is clearer, then identifying who should undertake what is generally more straightforward. This section, thus focuses on the key functions that institutions must undertake or enable (Table 13). An indication of the most appropriate level (or levels) at which the corresponding function should be promoted and implemented (within the institutional framework introduced previously) is also given in Table 13. The functions listed should not be reviewed as mandatory, but more as a check list.

An underlying need behind many of the key functions is that of education. The characteristics of groundwater resources are often poorly understood by policy makers and water users alike. Moreover, the social, economic, political and institutional factors governing groundwater use and the effectiveness of different institutional arrangements for resource management may not be adequately appreciated by policy makers and technical specialists.

*Resources Evaluation*

A realistic assessment of the status of groundwater utilization and resources provides the essential background against which the need for, and focus of, groundwater management activities can be judged. This assessment will vary widely in its degree of sophistication from preliminary evaluation of the groundwater balance and state of storage reserves (in cases where only reconnaissance data are available) to detailed numerical aquifer modeling (in cases where the necessary input parameters can be reasonably estimated and adequate water level monitoring is available for calibration).

**Table 13: Summary of groundwater resource management functions**

| Key functions | Potential activities | Institutional roles: NPM | RRB | AMC | WUA |
|---|---|---|---|---|---|
| Resource Evaluation | • assessment of status of groundwater resource exploitation (including use of numerical models) | pro | (imp)/inv | inv | inf |
| | • targeted monitoring of groundwater levels and quality | | pro/imp | imp/inv | inf |
| Strategic Planning and Coordination | • integrated analysis of socioeconomic roles/interactions of groundwater | pro/imp | inv | inv | inf |
| | • coordination with government/private sector institutions directly/indirectly relating to groundwater | pro/imp | inv | inv | inf |
| Identification of Management Priorities | • assessment of susceptibility to degradation | pro | imp | inv | inf |
| | • identification of resource conservation zones | | pro/imp | inv | inf |
| | • groundwater valuation and pricing review | pro | imp | inv | inf |
| Resource Regulation | • establishment/consolidation of register of abstractors, abstraction rights and water charges/markets | | pro/imp | inv | inv |
| | • water (re)allocation and dispute resolution | | pro | pro/imp | inv |
| | • demand management support | pro | pro | pro/imp | imp |
| | • compliance monitoring and enforcement measures | | pro/imp | imp | inv |

NPM: national planning ministry
RRB: regionally-based regulatory body
AMC: aquifer management committee
WUA: water-user associations
pro: promote
imp: implement
inv: involve
inf: inform

*Note:* The need for a tiered institutional framework will be evident from the respective roles identified, although it is not the intention to imply that a top-down as such is preferable; it may be better for components of the resources evaluation function not to be under regulatory agency leadership, since this may compromise their credibility in the eyes of water-users, although this will often prove difficult to avoid in practice.

***Groundwater Recharge.*** Quantification of the current rate of *groundwater recharge* to an aquifer is one basic prerequisite for efficient resources management. Groundwater recharge may be defined in a general sense as the downward flow of water that reaches the water table and forms an addition to the groundwater reservoir. A clear distinction should be made between the potential recharge from the soil zone and the actual recharge to the aquifers. These quantities may differ significantly due to interception by deep-rooted vegetation or by perched water tables.

At any location two distinct components of natural aquifer recharge can be recognized:

- *Direct (or diffuse) recharge* from rainfall (excess to soil moisture deficits and short- term vegetation requirements) which infiltrates directly
- *Indirect (or localized) recharge*, resulting from infiltration through the beds of perennial and ephemeral surface watercourses, and other forms of runoff.

In practice, a spectrum of processes between these two end members occurs (Lerner and others, 1990; Simmers and others, 1997).

In Figure 15 a broader conceptualization of groundwater recharge processes is introduced, distinguishing those that occur locally within the rural development area underlain by an aquifer system (be they as a result of natural direct/indirect processes or artificial causes) from those that occur at greater distance, especially in situations where the aquifer system extends into neighboring hilly terrain.

**Figure 15: Schematic representation and classification of aquifer recharge and discharge processes**

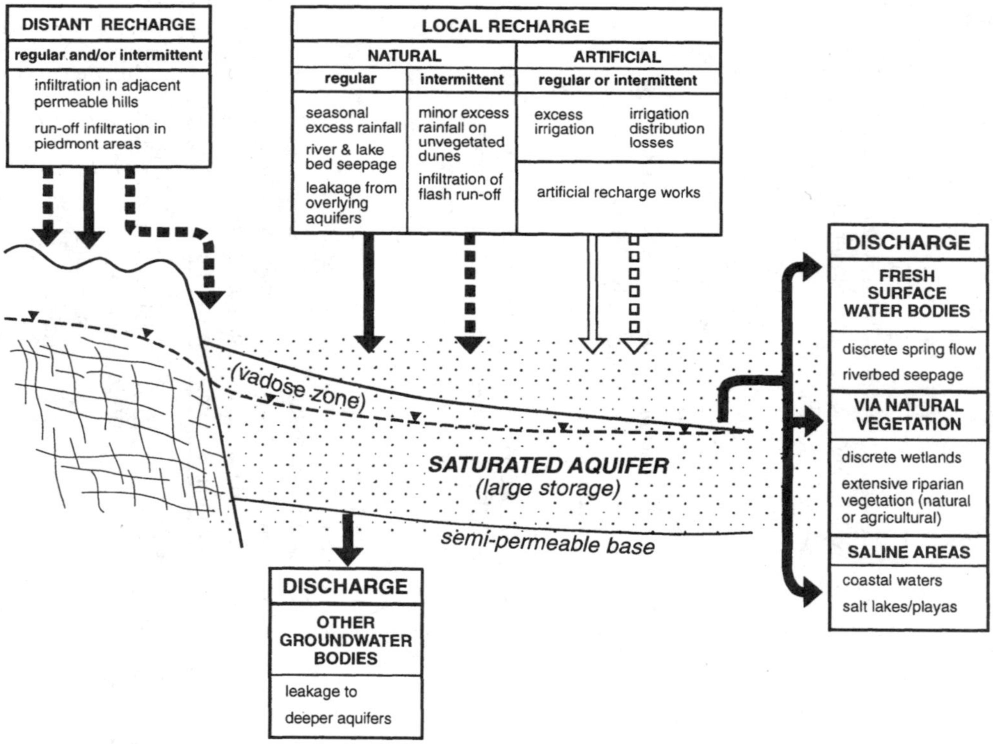


*Note:* Artificial discharge through pumping wells is omitted for simplicity; other geological structures will change the distribution and scale of recharge and discharge components.

A number of general observations can be made in relation to aquifer recharge:

- There is no doubt that recharge occurs, to some extent, even in the most arid regions, although areas of increasing aridity will be characterized by much decreased downward flux to the water table of much greater temporal variability (Figure 16).
- As aridity increases, direct recharge will become less important in terms of total replenishment than indirect recharge, and the artificial (or incidental) components of recharge arising from human activity also become increasingly significant.
- Estimates of direct recharge are always likely to be more reliable than those of indirect recharge.

For most practicable purposes it would be sufficient initially to estimate recharge rates in the more arid regions within two of the scale increments indicated in Figure 16, but even this will sometimes prove difficult. More precision can only be achieved through the analysis of carefully-monitored aquifer response to significant medium-term abstraction. The actual frequency of infiltration events, and the vadose (unsaturated) zone transit time until recharge reaches the water table, are also important considerations.

**Figure 16: Categorization of aquifer recharge in the more arid regions for practical groundwater resource evaluation and development**

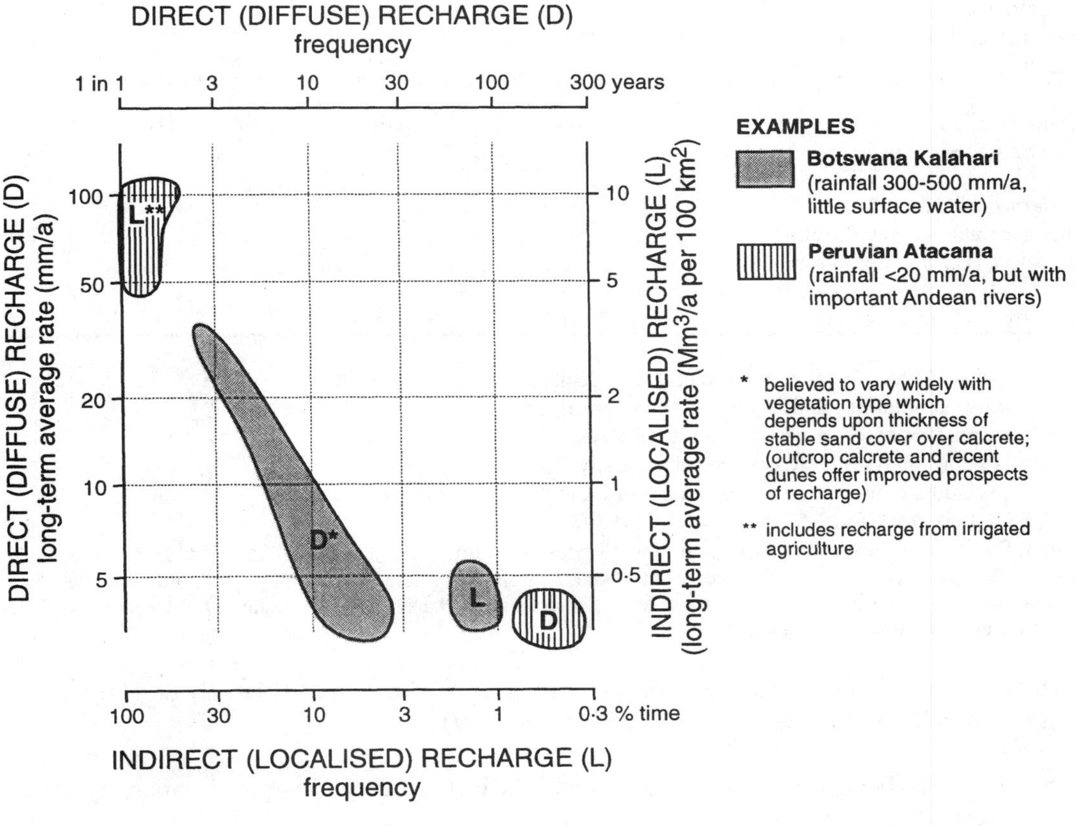


*Note:* In the examples shown, the Peruvian Atacama desert is a hyperarid region that fringes a major mountain chain from which numerous perennial rivers flow, whereas the Botswana Kalahari desert, although receiving considerably higher rainfall, is an extensive sand-covered plain covered with well established deep-rooted vegetation with significant soil infiltration but very low rates of groundwater recharge

*Source:* Foster, 1987.

Major, but very infrequent, recharge is a totally different proposition in resource management terms to more regular, if smaller, replenishment. This is because the negative side effects of excessive abstraction (albeit temporary) may have already occurred prior to replenishment. Further, the existence of an aquifer hydraulic gradient is no guarantee of recent recharge, since there may be "fossil gradients" reflecting historical recharge from past periods of much wetter climate, with natural recession of groundwater levels continuing to the present day.

The *quantification of groundwater recharge* is fraught with uncertainty and it is necessary to apply and compare a number of independent approaches (Foster, 1987). The main techniques that can be employed specifically to estimate current groundwater recharge rates may be divided into those for which the required data are often available or can readily be collected, and those for which more specialized and expensive facilities are needed (Table 14).

**Table 14: Principal direct techniques used for groundwater recharge estimation**

| *Technique* | *Applicability* | *Typical costs* | *Specialist needs* | *Time step* |
|---|---|---|---|---|
| *Conventional Methods* | | | | |
| Hydrometerological Data Processing (soil water balance) | D(L)0 | c** | • | ESYH |
| Hydrological Data Interpretation | | | | |
| - water table fluctuations | D(L) | c** | • | YH |
| - differential stream/canal flow | L | c** | • | I/E |
| Chemical and Isotopic Analyses from Saturated Zone | D+L | c-b* | • (•) | HG |
| *Modern Techniques* | | | | |
| Chemical and Isotopic Profiling of Vadose (Unsaturated) Zone | D0# | b-a* | • | HG |
| Soil Physics Measurements | D0 | a | • | SY |

D/L diffuse (direct)/localized (indirect) recharge distribution
0 only suitable for relatively uniform soil profiles
# inappropriate for irrigated agricultural areas
* isotopic analyses increase cost substantially
** excluding construction and operation of basic data collection network
a = >US$50,000; b = US$10-50,000; c = <US$10,000
E-event, S-season, Y-year, H-hydrological time, G-geological time
*Note:* The costs given are for a typical area of about 1,000 km$^2$; it should be noted that not all methods are appropriate for all conditions, and a clear conceptual model of possible aquifer recharge mechanisms is a pre-requisite for selection of technique and for quantitative evaluations based on field data

The applicability and potential accuracy of any given method depends largely on two semi-independent facets of the ambient conditions (Simmers and others, 1997):

- The superficial geological environment, which determines the spatial variability of the recharge process and the extent of development of surface runoff
- The vegetation system, whether native or agricultural and with or without irrigation.

Groundwater recharge estimates from individual techniques will nearly always be subject to considerable error. This will be evident from:

- Numerous limitations of each of the techniques described
- The wide spatial variability characteristic of rainfall and runoff events, especially in the more arid regions
- Widespread lack of lateral uniformity in soil profiles and hydrogeological conditions
- Frequent inadequacies in hydrogeological databases, especially in developing nations.

Moreover, the necessary database improvement to achieve more reliable recharge estimates will be more time-consuming than those required to improve all other factors affecting groundwater resources evaluation, especially if precise results are needed.

There is also a need to break the widely used paradigm that average groundwater recharge is constant with time. The evaluation of groundwater resources needs to identify linkages with land-use and surface water, especially in arid regions, where a major proportion of the total recharge may be derived from irrigated canals and/or irrigated fields (Figure 15). Modifications to canal construction and operation, irrigation technology and cropping regimes can then cause radical changes in groundwater recharge rates. Moreover, in some hydrogeological conditions major potential infiltration is rejected because of shallow water table, and recharge can be increased as a result of pumping for irrigation (Figure 17).

***Coping with Hydrogeological Uncertainty.*** When confronted by such uncertainty, it is strongly advisable for project design or management strategy to be sufficiently flexible as not to require radical change in the event of initial predictions proving subject to considerable error. Project design or policy formulation to accommodate initial uncertainty about groundwater recharge estimates is greatly aided by the use of *numerical (distributed-parameter) aquifer models.* These can be used to analyze critically the sensitivity of aquifer response to abstraction to errors in key parameters, such as groundwater recharge rate (Figure 18).

If sufficient groundwater level data (from an appropriate observation borehole network) corresponding to a period of significant medium-term abstraction, together with sound information of aquifer properties, hydraulic boundaries and groundwater discharge are available, then such a model could be used to determine historical groundwater recharge. In essence the technique involves varying aquifer recharge rates and distributions in the model so as to achieve calibration with historic groundwater level data, but it is very important that the sensitivity of such calibration to variation in recharge estimates be fully tested. A limitation of the method is the fact that the calibration achieved may not be unique. It may prove impossible to distinguish variations in recharge rates from those of other parameters, such as unconfined storage coefficient.

The value of an integrated operational approach to refining groundwater resource evaluations cannot be overstated. In this context it is vital that sufficient effort goes into monitoring aquifer response. Short-term economies in this respect are likely to prove counterproductive in the long run. In areas of complex hydrogeology, the approach (Figure 18) will be the only practicable way to improve the reliability of groundwater recharge estimates and of assessment of potential aquifer depletion-related side effects, and in many less complex situations it will often still be the most cost-effective way.

***Role of Groundwater Storage and Discharge.*** It is also of equal importance to elaborate a *realistic conceptual model* and quantitative estimate of the mechanisms and rates of *aquifer discharge* (Figure 15). This, inevitably, is an aquifer specific activity, but the results provide a cross-check on recharge estimates, and can also reveal key linkages to elements of the surface water environment which are dependent upon the groundwater flux of the aquifer system concerned. It is important to distinguish:

- Discharge to freshwater systems, since these may be required to sustain downstream uses, for water supply interests and/or other river interests
- Discharge via natural vegetation, including sustaining ecologically and/or economically valuable freshwater wetlands and brackish lagoons
- Discharge to saline areas including coastal waters, salt lakes and playas.

**Figure 17: Increasing groundwater recharge to the shallow alluvial-deltaic aquifer or Bangladesh by controlled water table lowering**

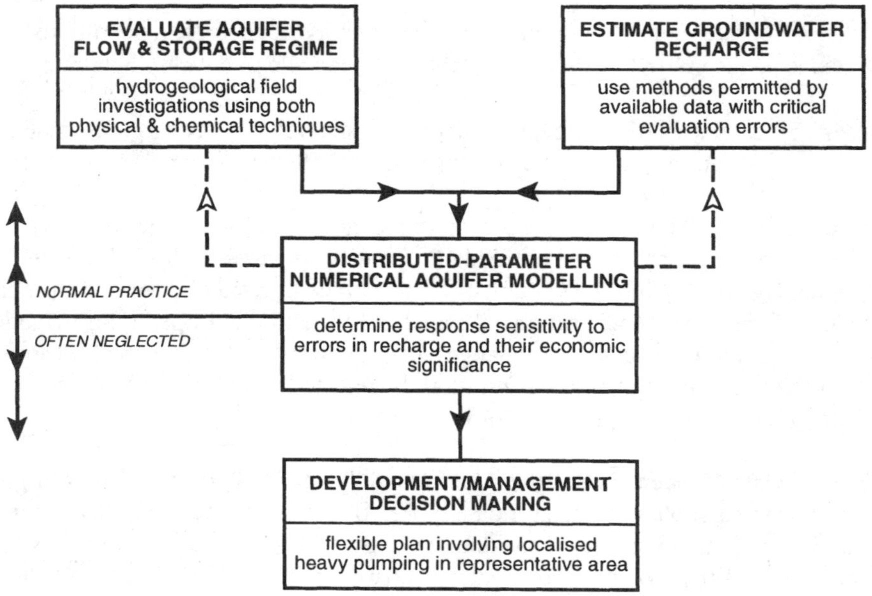


*Note:* An average of 320 mm/a of potential recharge is estimated to be rejected at this location, because of water table surfacing in the wet season, and the figure illustrates how recharge could be increased with pumping for agricultural irrigation in the preceding dry season.

**Figure 18: Organization scheme for use of numerical aquifer modeling to inform groundwater management plans**

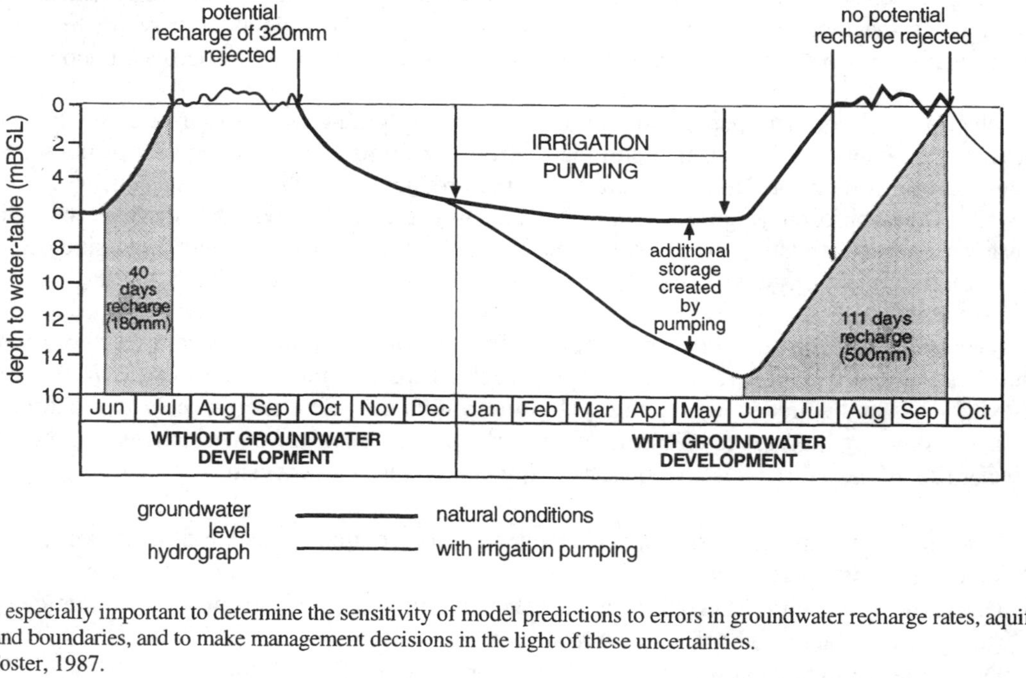


*Note:* It is especially important to determine the sensitivity of model predictions to errors in groundwater recharge rates, aquifer storage capacity and boundaries, and to make management decisions in the light of these uncertainties.
*Source:* Foster, 1987.

Since all groundwater flux in an aquifer must be discharged somewhere, the question of the *safe yield for groundwater exploitation* arises (Bredehoeft, 1997; Sophocleous, 1997). This should be recognized as an essentially subjective concept. Safe yield is obviously bounded at the upper level by the long-term active aquifer recharge rate. However, it should (but all too frequently does not) involve value judgements about the importance of maintaining (at least a proportion of) some of the discharges from the aquifer system. This is not straightforward, but it is obviously essential that the groundwater resource evaluation process should at least identify all downstream linkages and dependencies.

Another key aspect of groundwater resource evaluation is assessing the *volume of exploitable aquifer storage* and the *susceptibility of the aquifer system to adverse effects* if subjected to either short-term (temporary) or long-term overdraft due to excessive pumping. It is important for cost-effective groundwater and land management to diagnose this susceptibility adequately. In many ways the vast storage of many groundwater systems is their most valuable property, and this needs to be exploited in a strategic fashion. The key question is how to use (but not to abuse) this storage resource. In some cases the social, environmental and economic value of the services provided by maintaining groundwater levels close to their natural fluctuation should logically preclude the full exploitation of groundwater storage. However, in others it should be possible to physically engineer or economically compensate for impacts on groundwater-level-related services, allowing a fuller development of deeper groundwater storage.

Groundwater monitoring and data collection can be costly and time consuming. Information requirements must always be carefully targeted towards management decisions, but long-term sustainability issues should not be lost sight of. It is also extremely important that synoptic data of aquifer evaluation and groundwater resource status are regularly and systematically disseminated in a suitable format to the principal stakeholders and the general public.

### *Resource Valuation*

The value of groundwater can vary from next-to-nothing to being priceless. It is next-to-nothing in a deep aquifer beneath a remote sparsely-populated area with abundant surface water resources. It is almost priceless in a wellfield which is the sole source of drinking water for a town in the desert. Most "real world cases" lie between these two extremes. Valuations are urgently needed to inform the development and implementation of management policies for groundwater resources (Young, 1996). But how can the value of water actually be assessed?

A clear distinction must be made from the outset between the *benefits of using water* (either today or in the future) which determine its value, from the *costs of providing the water supply*, which is a different concept . If this cost is high it could be an indicator of high value, but it is equally possible that the costs of water supply exceed the benefits of water use and that the water is merely misallocated. Logically groundwater abstraction should only be undertaken if the net benefits (defined as the benefits from water use *minus* those that would have accrued by conserving its natural state *minus* the water supply costs) are positive (Schiffler, 1998).

A last distinction that needs to be made in valuing water is between the benefits of a *unit of water supplied* and the benefits of a *unit of water stored*. Usually the valuation of water is applied to the value of a unit of water-supplied. The same valuation can be applied to future use. By discounting the value of future flows and dividing it by the present water stock, the value of a unit of water stored can be estimated. This method allows estimation of the option value of using water stocks in the future rather than now.

A further difficulty in groundwater resource valuation is *seasonal variation in the value of water*. In many cases, during the wet season groundwater is usually abstracted for irrigation in only small quantities, because of the availability of other water sources. The value of groundwater at that time is always close to zero. In contrast during the dry season, the value of groundwater can vary depending on crop demands and weather conditions. These variations can actually be observed where well owners sell water on a daily basis to farmers in areas without adequate water supply.

Valuations obtained directly from competitive market transfers (the *revealed preference method*) can be relied upon to assess the benefits derived and thus the economic value of most other goods, but are rarely

**Table 15: Summary of economic methods applied to groundwater valuation**

| *Valuation* | | | |
|---|---|---|---|
| *Method* | *Approach* | *Applicability* | *Limitations* |
| Revealed Preference | direct observation of how much water users actually pay to water vendors or in water markets providing supply competitively | • extractive value for domestic water supply or irrigated agriculture as appropriate | • gives upper limiting value if limited quantities traded, since cannot be assumed that insensitive to quantity provided |
| Water Productivity Residual Value | traded value of goods measured by the gross margin (turnover net of taxes minus material inputs) divided by amount of water used | • extractive value in irrigated agriculture (data widely available) | • method discredited because often erroneously assumes all benefits imputed to water (ignoring land, labor, etc.)<br>• can be distorted by substitution of capital investment for water-use |
| Water Rent Residual Value | as for water productivity but also subtracting cost of labor, capital and land, reflecting more closely true opportunity cost | • extractive value in irrigated agriculture (additional data not so readily available) | • measurement and valuation of family labor difficult<br>• hypothetical rental value of land (without water) problematic<br>• interest rate applied to capital is subjective and discretionary |
| Land Price | price differential between identical land plots with/without groundwater source multiplied by market interest rate divided by annual water use on plot concerned | • extractive value in irrigated agriculture | • various other characteristics (soil, topography, access, etc.) affect value of land<br>• too few land transaction prices may be available |
| Contingent Valuation | survey by interview of willingness to pay for given water supply or preservation of groundwater-fed ecosystem | • extractive value for domestic water supply or in-situ value for maintaining ecosystems | • still controversial, since prone to bias according to interviewees expectation/knowledge income-level/gender |

*Note:* It is evident that the application of some methods may be seriously questioned as a result of excessive simplification and/or inadequate data availability for their optimum application.

available for water. A number of alternative techniques (Table 15) are being employed by economists for evaluation of groundwater (NRC, 1997; Schiffler, 1998).

***Extractive versus In-Situ Values.*** In the case of groundwater resources it is not sufficient to estimate the extractive value of the groundwater for use in a given economic activity alone, but an estimate of the forgone benefits resulting from the groundwater not having been maintained in-situ in the aquifer are also extremely important. In some instances, the *in-situ value* can be calculated directly by an analysis of the environmental damages resulting from groundwater abstraction, such as those caused by land subsidence. However, while these damages can be estimated in a relatively straightforward manner ex-post, it is considerably more difficult ex-ante, because of hydrogeological uncertainty in the aquifer response to groundwater depletion.

Other components of the in-situ value can in some circumstances be obtained by the *residual value and land price differential methods* (Table 15). For example, if it is estimated that a portion of a coastal aquifer could become completely unusable through saline intrusion for a period of 10 years, and subsequently its water could only be used to irrigate a very limited range of crops for a further 20 years, then the in-situ value of the groundwater can be estimated by comparing the net present value of the water rent over the period concerned with the corresponding value over the same period of time from an undamaged aquifer.

***Average versus Marginal Benefits/Costs.*** *Average benefits/costs* are determined by estimating the total benefits/costs associated with a water abstraction during a specific time period and dividing them by the total amount of water abstracted in the same period. This averaging approach can, however, be misleading. On the *benefit side*, if only a limited amount of water is available, it is normally used for high-value uses, while as more water becomes available, it is used for lower-value uses. The more meaningful concept is of *marginal benefits/costs*, where only the benefits of additional water made available (or the loss of benefits incurred if water supply is reduced) would be considered. The marginal costs differ widely according to whether additional capacity is needed or not. If spare capacity exists with existing installations, the marginal costs are limited to additional energy to run pumps and are often referred to as the *short-run marginal costs*. If, however, additional capacity is needed, the marginal costs increase sharply (because of the capital cost of the new infrastructure) and are then referred to as *long-run marginal costs*.

Capital costs usually account for a large share (50-90 percent) of total long-run marginal (incremental) costs of water supply. These capital costs can be estimated in two different ways:

- Total investment cost can be multiplied by a cost recovery factor (including depreciation period and interest rate) to determine the annual capital costs; the annual capital cost is then divided by the average amount of water abstracted annually to yield the capital cost per unit of water
- A dynamic cost calculation, which takes into account the fact that the capacity of infrastructure may not be fully used in the initial period by considering the expected stream of investment costs and water abstraction, which are then discounted to present.

Both methods are very sensitive to the interest (discount) rate chosen.

***Financial versus Economic Benefits/Costs.*** Both the benefits and costs of using water can be estimated either in *financial* or in *economic* terms. In financial terms, all prices are expressed in market prices actually paid, while in economic analysis prices are expressed in shadow terms, which reflect actual scarcity value after excluding any so-called transfer payments (such as taxes, import duties, fees and subsidies) and correcting for any price distortions (arising from price controls or overvalued exchange rates).

Using *financial costs*, which for many non-economists seems to be the easiest way to proceed, can lead to a serious misallocation of resources. For example, consider a country which subsidizes energy prices and the cost of capital while it supports agricultural prices. By doing so it lowers the financial costs of a water abstraction and increases the financial benefits from irrigation. A financial analysis of a water abstraction/supply may show that there are net benefits but an *economic analysis*, which excludes these price distortions, may show that it is more beneficial to keep groundwater in its natural state or that net benefits are higher if water is used for other purposes at a different location.

If water use is consumptive (incorporated in a product, transpired or evaporates) the *benefits/costs of return flows* can safely be excluded from economic analysis. In many cases, however, part of the water supply is not used consumptively, but returned to the environment as domestic effluent or agricultural drainage. If these return flows are significant, their value has to be assessed; they can be positive but are often negative as a result of quality deterioration during use.

*Strategic Planning*

Because groundwater resources management options are affected by a wide variety of technical, social and economic considerations, *integrated analysis* and strategic planning is a key function providing a foundation on which progress can be made. Moreover, it is essential to *coordinate actions* with those responsible for policies affecting energy tariffs, crop prices, fertilizers and pesticides subsidies, and so forth, since interventions here can have a major impact on groundwater resources. Another key aspect will be macro-planning on the scope for and implementation of conjunctive use of surface and groundwater resources.

At the regional level it is extremely important to undertake clear *identification of priorities*. This process will include:

- The assessment of the susceptibility of groundwater systems to degradation through inadequately controlled exploitation which (along with resource valuation) condition the priority for action
- The identification of especially critical areas on the above basis and their declaration as *resource conservation zones*, a concept which can play a critical role in the development of effective groundwater management.

Such zones may not need to address entire aquifer systems, since it is important to recognize that many problems can be addressed successfully at more local scale. While their boundaries need to be consistent with the hydrogeological regime, it will be necessary from a political perspective to reflect social or administrative boundaries, since these may well effect the viability of management actions. These zones are also important from a common-property perspective, since they define the boundaries within which management actions will be taken, and help to limit the array of water users and other stakeholders that need to be involved.

Equally important for the clarification of management priorities is deciding what are the *key services provided by groundwater resources* (such as irrigation of certain types of crop, provision of domestic water supply, environmental discharges, and so forth), because this is what society cares about. In some ways this forms part of resource valuation, but inevitably on a qualitative basis. In other ways it is a social exercise to clarify the wider objectives that society wishes to achieve through groundwater management and is therefore, a key prerequisite to shaping regulatory targets.

*Resource Regulation*

The regulation of groundwater resources is a many-facetted process (Salman, 1999) which is best carried out on a flexible and adaptive basis though the collaborative efforts of some form of local regulatory agency/authority, aquifer management committees and local water-user associations. Amongst the key activities which may be needed are the establishment/consolidation of a *register of abstractors* and the organization of water abstraction rights, together with the establishment of some form of abstraction charging or water markets.

Regulatory functions are central to groundwater management in all situations where the characteristics of the groundwater resource are not such as to be effectively self-regulating (Figure 19), and especially where the risks of irreversible degradation are significant (Tables 11 and 12). The nature of this function and the institutional level at which it needs to occur, however, can vary considerably. In some cases, regulation will be a high-level function governing the operation of the water-rights system and the operation of water markets. In others, it may be a more localized procedure such as codes of practice on permissible well spacing or irrigation technology (Box 8).

Direct regulation of groundwater use is often extremely difficult because of the large number of geographically dispersed abstraction points involved. Additionally, unless broad social support exists for regulation, enforcement is often politically sensitive and problematic. Thus while a broad regulatory framework is required to provide the platform on which other management approaches operate, it is rarely effective on its own.

***Definition of Groundwater Abstraction Rights.*** A form of rights system related to groundwater abstraction and use is operative in various nations. In some cases, however, such rights are only informally established on the basis of social practice, whereas in others they are formally registered and encoded in law. In many situations, the clarification of groundwater rights (and in certain instances rights reform) will be the essential prerequisite for introducing resource management measures. It is important to emphasize that water rights systems are not inherently dependent on government agencies or formalized legal systems, and can be carried out through social processes. Where feasible, active self-governance is (in the long run) preferable to the imposition of government rules.

Once national legislation exists for the definition of groundwater abstraction rights, either on a universal basis or in specific areas of concern, the key initial process will be to consider the claims of existing abstractors. The administrative process is likely to be protracted and may encounter a number of significant problems, which need to be planned for and confronted systematically if the process is not to be baulked.

First, how to cope with pre-existing unauthorized abstractors, in situations where some form of licensing had previously been in operation. A parallel problem arises in situations where existing authorized abstractors make claim for much larger volumes than appear reasonable from observed actual or probable historic use. In both cases, the formalization of water rights normally involves recognition of illegal abstractors, provided that an appropriate claim is made during some form of "truce period" and that the usage is adequately justified. It should be noted, however, that this process may be strongly resented by existing authorized abstractors (who have conformed with past regulatory provisions) and some form of initial financial concession in relation to the charging for abstraction permits would appear appropriate.

Second, how to cope with a large number of small abstractors. These are usually exonerated from any payments for abstraction permits, but their existence and rights need to be clearly registered, so that they can be appropriately protected in the future. Where small irrigation wells are involved the preferred option is to group their claims for abstraction rights under a single title, held by some form of water-user association or irrigation committee.

Third, how to cope with the fact that the total claimed groundwater abstraction rights by existing users may considerably exceed current estimates of the available resource. The sensible approach here is to offer some form of time-limited abstraction right, whose volumetric entitlement is reviewed every 5-10 years and can be reduced in the light of new understanding of (or physical processes affecting) the availability of groundwater resources.

In any situation of water scarcity, an adequate process for allocation of available groundwater resources among competing uses will be a critical task. While this can be achieved directly through the definition and revision of water rights, it may be more easily tackled through the establishment of water rights markets.

Since groundwater is often of fundamental importance to life, health and livelihood, disputes over management interventions are likely to be relatively common, and a transparent mechanism for dispute resolution is a key management function. Disputes, however, may be avoided by the provision of technical and financial support for water users in relation to the introduction of demand management measures, which while making real water-use savings promote higher water-use productivity. Public relations are also extremely important, and both water users and the general public need to be kept informed about the state of groundwater resources and the benefits of sound resource administration.

## *BOX 8: Capacity for Indirect Regulation of Groundwater Abstraction in Bangladesh*

• Bangladesh has an urgent need to augment foodgrain production and to alleviate rural poverty. More than 80% of the total population (of about 110 million) rely on agriculture for subsistence and employment. The availability of land in the dry season means that irrigation from groundwater is economically attractive and provides the quickest route to raising agricultural production and rural employment.

• Groundwater development for irrigation commenced in the early 1970s and reached an area of 7,300 $km^2$ by 1985. Initially it was financed and operated by public sector institutions using high-yielding deep tubewells, which provided water to rural land owners, but from 1975 there was a major increase in privately-owned low-cost shallow tubewells equipped with surface suction-lift pumps powered by diesel engines. By 1985 it is estimated that 173,500 of these were operating, together with 285,400 manually-operated shallow tubewells, compared to only 17,200 deep tubewells, since farmers prefer the smaller units which involve less dependence on water purchase from deep tubewell operators.

• However, shallow tubewells with suction-lift pumps can only lift water from depths of 6-8m (Inset). The absence of a significant monsoon in 1983 (believed to be a 1-in-25 year event) led to failure of shallow tubewells in some areas and raised doubts about the appropriateness of this technology, (Pitman in Kahnert & Levin, 1993).

• Under natural (pre-development) conditions the complex layered alluvial-deltaic aquifer of Bangladesh has groundwater levels virtually at surface during the wet season (July-October) with a recession down to 3 m depth by the end of the dry season (May). This means that much potential recharge was rejected because the aquifer was full for a substantial part of the wet season. Infiltration of rainfall to groundwater can be increased from 150-200 mm/a to 400-500 mm/a if the water-table is lowered by abstraction for irrigation.

• The degree of water-table lowering will depend on the proportion of the land area irrigated in the dry season and the specific yield of the surficial strata experiencing drainage. In areas where the latter is high, it is possible to maximise (and to access) all potential recharge with shallow tubewells powered by suction-lift pumps. In such areas deep tubewells should be discouraged as being less economic and potentially conflictive. However, in other areas deep tubewells with lineshaft pumps are required and the water-table may be drawndown to 10-15 m by the end of the dry season, which can cause serious interference with shallow tubewells. Modified shallow tubewells with lineshaft pumps, capable of yielding 5-10 l/s from up to 11 m depth have also been developed for these areas.

• Policy reforms were also being implemented in the late 1980s to deregulate the importation and purchase of tubewell pumps, engines and fertilisers for groundwater irrigation, so as to stimulate further development of privately-owned shallow tubewells. Moreover, it is hoped that improved hydrogeological knowledge of the storage properties and dynamic water-table fluctuations of the alluvial-deltaic aquifers will help refine the mapping of areas appropriate for groundwater exploitation by the different tubewell technologies.

| TYPE | | Hand-Pump Tubewell | Surface-Suction Tubewell | Deep-Set Suction Tubewell | Lineshaft Deep Tubewell | Modified Shallow Tubewell |
|---|---|---|---|---|---|---|
| Borehole | Type | shallow tubewell | shallow tubewell | shallow tubewell | deep tubewell | shallow tubewell |
| | Depth | 30-35 m | 40-60 m | 40-60 m | 100 m | 40-60 m |
| Pump Unit | Type | surface mounted suction-lift | surface mounted suction-lift | pit installed suction-lift | lineshaft turbine | lineshaft turbine |
| | Power | manual | diesel engine | diesel engine | diesel engine | diesel engine |
| Maximum Lift | | 7 m | 6 m | 8 m | 12-15 m | 11 m |
| Typical Yield | | <1 l/s | 1-2 l/s | 1-2 l/s | 40-60 l/s | 5-10 l/s |
| Maximum Irrigated Area | | 2 ha | 4 ha | 4 ha | 50 ha | 10-15 ha |
| No. of Farmers Using Water | | up to 10 | up to 20 | up to 20 | up to 200 | up to 50 |
| Construction & Operation Arrangements | | installed by public sector agencies operated commercially | installed/operated by private owners | installed/operated by private owners | supplied by public-sector agencies, sold/rented/operated by cooperatives | installed/operated by private owners |

***Inset :*** *Different types of water-supply borehole technology operating in Bangladesh with their yield and lift constraints*

***Licensing of New Production Boreholes.*** In the case of direct water administration by an agency of regional/local government, experience suggests that regulation of groundwater abstraction has been more successfully achieved by exercising *control over the construction of waterwells* themselves - their numbers, depths and diameters. In this situation the normal legal requirement would be that any individual or company wishing to drill a borehole, dig a well or capture a spring to exploit groundwater resources requires a *construction permit*. The detailed procedure will vary to some degree with the state-of-knowledge of groundwater resources in the given area.

It is helpful for the regulatory agency to be in a position to offer technical advice to the applicant, since this will build better relations and will ensure the return of reliable data on the well drilled. This technical advice should include opinions on:

- The maximum yield obtainable from the given aquifer in the area concerned
- An estimate of required well depth
- The appropriate separation of the new well from existing wells
- The well-screen required, the preferred construction materials and the wellhead sanitary protection.

The second stage is the issuing of a *license for abstraction* once the new installation is constructed. The regulatory agency would normally reach a decision on permissible yield on the basis of:

- The state of groundwater exploitation in the area
- The proposed use, including the area and method of proposed irrigation
- The quality requirements for the proposed use.

While it is desirable for all water supply installations to require a drilling permit, since this is the only way in which the regulatory agency can effectively control exploitation and avoid irrational development, many exonerate small-scale users from the need to obtain an abstraction license.

There are various ways by which yield may be controlled:

- The penetration of the borehole in the saturated zone of the aquifer
- The diameter of the borehole, and in effect the size of pump that can be installed
- The type of pump installed
- The hours of pumping per day, and annual or monthly abstraction rates (by direct metering or indirect estimation).

Abstraction rights and regulations have little meaning unless they relate to actual practice on the ground. Some form of periodic inspection in the case of large abstractions (and occasional spot-checks in case of smaller ones) are necessary to *enforce abstraction control policy*. Monitoring of compliance is an important function, which can take place either in a formalized way by the regulatory agency or (as is the case with most traditional water rights systems) at the community level. The related enforcement function can be achieved through a variety of informal and formal social processes, with police powers being only one (relatively limited) avenue.

***Sanctions for Non-Compliance.*** For any policy to be effective some form of legal penalties against those who construct waterwells without permit or exceed the licensed abstraction are required. These normally include such actions as temporary (or even permanent) prohibition on the use of the well, depending on the scale of the offence and its effect on third parties or on the aquifer resource itself. Monetary fines may be considered but these are not normally considered appropriate.

**Figure 19: Variation of groundwater resource regulation requirement with hydrogeologic setting and socioeconomic circumstances**

| SOCIOECONOMIC AND HYDROGEOLOGIC CONTROLLING FACTORS | INCREASING NEED FOR RESOURCE REGULATION | | |
|---|---|---|---|
| | NATURAL REGULATION | PARTIAL CONTROLS | FULL REGULATION* |
| no. of groundwater users | small → | → | large |
| level of water demand | low → | → | high |
| cost of alternative water-supplies | low → | → | high |
| climatic regime | humid → | → | arid |
| functions/value of shallow aquifer | minor → | → | major |
| aquifer susceptibility to degradation | low → | → | high |
| aquifer T/S ratio** | small → | → | large |
| aquifer $S_S$/R ratio*** | small → | → | large |
| **VARIATION WITH MAIN AQUIFER TYPES** (see Table 3 for key) | ← WCB → | | |
| | ← IMB → | → | |
| | | ← CSA → | → |
| | | ← RCL → | → |
| | | ← MAF → | → |

* full regulation includes water rights allocation, water-user participation and economic instruments

** aquifer response characteristic or ratio of transmissivity and storativity (can vary in large range from$10^{+1}$ m/d to $10^{+8}$ m/d)

*** aquifer storage characteristic or ratio of specific storage and average recharge (can vary in range $10^0$ to $10^{+5}$)

*Note:* Only a qualitative indication of priorities is possible since local circumstances will exert important influence. It also should be noted that well registers and general construction guidelines are recommended in all situations.

## Economic Instruments for Groundwater Management

There is an array of economic instruments for groundwater management, among which well licenses and abstraction fees are best known. However, there are a number of policy issues in other sectors which can have more pronounced impact on groundwater abstraction, but which are seldom considered as instruments for groundwater management. Among these are energy tariffs, import restrictions and duties for agricultural products, subsidies for drilling wells and buying pumps or to purchase water-saving technology.

The most appropriate instrument depends partly on the local social and institutional situation. Wherever water users own wells and pumps (either individually or collectively) they should generally bear the full costs of operating, maintaining and replacing them. Moreover, governments need to consider levying a resource management charge, which can also reflect resource scarcity and external impacts.

### *Abstraction Charges*

*Groundwater abstraction charges* in the form of a volumetric charge on actual abstraction (as opposed to a fee whose level is based on licensed abstraction) are not very common throughout the world. Since the 1960s they have been mainly introduced in industrialized countries with a humid climate, where groundwater is mostly used for municipal and industrial uses. Notably, in the few developing countries which have introduced groundwater abstraction charges (Jordan, Mexico, China and India), agriculture (which is by far the largest groundwater user) is still exempt from (or pays nominal) charges.

***Control of Revenues.*** If the revenues from groundwater abstraction charges go to the general budget of central or local government, there will be temptation to use the charge for fiscal purposes. It is far preferable to use this revenue to cover the *administrative costs of regional water-management agencies and/or local water-user associations*. If these entities are under the scrutiny of their constituents and act in a transparent manner, the revenues are likely to be spent efficiently and will not be an undue burden for water users.

If the charges are not sufficiently high to constrain groundwater abstraction, they should be increased beyond the level necessary to cover administrative costs. Water users should, however, have a say in the use of surpluses generated from these charges. One possibility would be to subsidize the purchase of water-saving equipment.

***Structure and Enforcement of Charges.*** An advantage of using abstraction charges to reduce exploitation (instead of using *license abstraction limits*) is that charges achieve the objective at minimum cost, by giving incentives to farmers to undertake water savings where they can be achieved. This leads to lower costs for all farmers than under a regime of abstraction limits.

*Volumetric groundwater abstraction charges* can be structured in various ways, but in reality charge scales are often linear. However, charges can also be progressive, with higher unit charges being levied for higher levels of abstraction, similar to increasing-block water tariffs commonly used in urban water supply in many developing countries. The charges can also be differentiated by type of use, although such a differentiation distorts incentives for saving water.

Another possibility is to levy higher charges during the dry season than wet season or higher charges for consumptive use than non-consumptive use, because in the latter case return flows are available for other uses. Such decisions depend on local hydrogeological conditions, since in some cases dry-season abstraction actually increases the storage capacity of the aquifer for recharge during the wet season.

In many countries it may prove difficult to monitor groundwater abstraction and to enforce abstraction charges. Most wells in rural areas in developing countries have no meters, and in those that do the meters are often broken or manipulated. To prevent this from happening, it is crucial that farmers understand the consequences of groundwater overpumping. It is equally important that they have a say in determining the objectives and instruments of groundwater management, including the level and structure of charges.

### *Indirect Influences*

***Energy Prices.*** Energy prices in developing countries are widely subsidized. In remote areas (without electrification) diesel pumps are still used to pump groundwater and diesel prices may be fixed at low levels. It may be politically difficult to increase the price of diesel, because of its impact on transport costs and the prices of many other goods.

In some countries, electricity tariffs for agricultural purposes are set at low levels, and sometimes flat-rate tariffs (independent of consumption) apply. Changing these tariffs could provide a major incentive for reducing groundwater pumping and economizing on water use. The level and the structure of *rural non-domestic electricity tariffs* might even be differentiated according to local groundwater resource status. However, in areas characterized by traditionally low electricity tariffs, any increase may be politically problematic.

***Import Liberalization.*** Import restrictions (such as bans or quotas) and import duties on agricultural products can keep the national price of these products above world market level. Overvalued exchange rates have a similar effect, although they also make fertilizer imports more expensive. High domestic prices for agricultural products are a strong incentive to increase production, often at the expense of groundwater resources. The exploitation of fossil aquifers in Saudi Arabia and some Gulf states for growing wheat illustrates the relationship between market protectionism and groundwater abstraction.

The trend towards *liberalization of agricultural markets* can be expected to continue, and trade will shift agricultural production from high-cost producers to low-cost producers. To the extent that the pricing of groundwater reflects its scarcity, agricultural production will shift from water-scarce areas to areas with rainfed production or with irrigation from more abundant water resources.

***Subsidized Credit.*** Subsidized credit has been widely used throughout the world to promote rural development. In some countries a large share of this credit has been used to drill deep wells and to buy turbine pumps and irrigation equipment. *Subsidized credit channeled through agricultural banks*, and often supported by international donors, has thus often exacerbated groundwater overexploitation. The problem is somewhat less virulent today, as many countries have phased-out subsidized credit, mainly because it has failed to reach the poor and has not been financially sustainable. The main current issue in rural finance is to give farmers access to formal credit at cost-covering interest rates. Such credits may be used to drill or deepen wells, but should preferably be invested in more efficient irrigation technology.

*Role of Water Markets*

*Informal water markets* are widespread in some developing nations. These markets generally involve local transactions between well owners and other users adjacent to each well. They function on the basis of informal (but socially accepted) agreements between individual well owners and those seeking to purchase water, and involve transfers of water already abstracted but not of water rights (Shah, 1993). These water markets are fundamentally different in scale and operation from those functioning formally on the basis of trade in legally defined rights to abstract a given volume of groundwater.

Both types of water market communicate a portion of the economic value of groundwater to both buyers and sellers. Within informal markets, however, this value is generally related to pumping costs, short-term availability and local use within agriculture. In contrast formal markets, functioning on the basis of a clearly-specified rights system, often communicate the difference in value between different uses and locations (such as public water supply versus agricultural irrigation) and bear some relationship to water availability in the larger sense of sustainable abstraction and aquifer storage.

Neither form of water market, however, reflects the *in-situ value* associated with groundwater or the third-party costs resulting from its exploitation. Informal water markets often enable different classes of users to obtain access to groundwater and serve to allocate access to (but not shares in ) the resource base, but often exacerbate overexploitation, since well owners pump as much water as they can in order to maximize returns from water sales. In contrast, *markets based on volumetric rights systems* can cap abstraction at sustainable levels and assist in allocating shares in the resource between use categories.

The technical, administrative and social aspects of rights definition pose a major difficulty for the introduction of satisfactory water markets in the groundwater case. First, groundwater systems are often poorly evaluated and monitored, and thus the quantitative basis essential for defining rights tends to be weak. Second, in some countries, the number of wells that would need to be monitored is extremely large, many being located remotely on private land (Moench, 1994), and thus there is a large cost and logistic difficulty inherent in actually monitoring groundwater abstraction. Third, water rights systems are socially complex and often based on deeply-embedded cultural values, and attempts to reform them can run counter to social characteristics.

Overall, it is important to recognize the role that water markets can play as part of the institutional framework for groundwater management. In many situations there may be no realistic alternative mechanism to achieve the needed re-allocation of resources. However, it is also important to recognize their limitations and that full collaboration of stakeholders through water-user associations or aquifer management committees will be needed to regulate their role. In many situations, water rights reform represents a major hurdle that can only be addressed over the long term. In addition, even if established on the basis of a clear rights system, markets will need to be regulated in order to address third-party impacts of groundwater abstraction and to allow for in-situ groundwater values.

## Mounting a Groundwater Management Action Plan

The operationalization of a groundwater management plan requires translating the general considerations outlined above into practice (Boxes 9 and 10). Although this task can be complicated and is to a significant degree dependent upon the adaptive capacity of local society (Turton, 1999), a phased implementation of different management measures will greatly facilitate the process. The discussion here assumes that a basic groundwater monitoring system is in place and that some government organization with at least limited regulatory power exists.

*Demand Versus Supply Side Measures*

From the outset it must be recognized that in the rural context there is generally much more scope for demand-side management measures (controlling irrigation groundwater abstraction) than there is for augmenting groundwater resources (through artificial aquifer recharge). The most significant irrigation water demand management measures will be those that lead to significant real-water savings through the reduction of non-beneficial evaporation and evapotranspiration, rather than those that improve irrigation efficiency through reduction of recirculation by deep infiltration to groundwater systems.

Substantial real-water savings can be achieved by preventing direct evaporation from irrigation water distribution systems, by reducing direct evaporation from open soil, by eliminating evapotranspiration by non-agricultural plants from perched water tables and by optimizing water use by the agricultural crop itself. This will require significant investment in improved water-distribution systems, irrigation technology, agricultural crop husbandry and in many instances changes in cropping regime with implications in terms of marketing agricultural produce. However, once the potential real-water savings are definitively identified and such investments made (probably through some form of loan provision and extension service), then it is likely that improvements in irrigation water productivity will be generated, and the overall result should not be a reduction in farmer"s income. However, it will be necessary to prevent use of the water saved simply to amplify the frontier of irrigated agriculture.

Although the scope for supply-side measures is generally more limited, the potential for integrated water resource management, and the need for more imaginative evaluation and development of conjunctive use should not be lost sight of. This type of approach (where physically feasible) is likely to result in greatly increased water supply availability during the dry season. However, there will often be significant institutional and operational barriers to be overcome to realize this potential.

*Slowing Growth in Abstraction*

Where monitoring data indicate that groundwater overdraft may be occurring or is imminent, *the first step is to limit the growth in abstraction*. This can give critical "breathing space" while more comprehensive management measures are implemented. Initial limitation of the growth in abstraction can be achieved through simple policy measures such as:

- ***Public education.*** A prerequisite—this should be focused on overdraft problems and their potential implications for water users. However, the information provided needs to be clear, consistent and closely correlated with the water-users experience.
- ***Imposition of drilling moratorium.*** This is often much more easily done than regulating abstraction itself, especially where drilling rigs (particularly those capable of drilling to considerable depth) are government operated. Even where they are not, the number of operators involved tends to be relatively small.

## *BOX 9 : Policy Options for Stabilising the Groundwater Resource Situation in Mexico*

• Some 75% of the total population (98 million), 70% of industrial development and 90% of irrigated agriculture in Mexico is concentrated in the northern and central regions, which have less than 20% of the national water resources. In these regions there is heavy dependence on groundwater for urban and industrial water-supply, and for agricultural irrigation.

• The Comision Nacional del Agua (CNA) have declared more than 100 aquifers as seriously overexploited. In some areas groundwater levels have fallen by 5m/a, and well depths and pumping costs have increased many fold since 1970. Agriculture is the largest groundwater abstractor, but there is serious competition for available resources around the larger urban and industrial centres. Contamination through induced infiltration of uncontrolled urban and industrial effluents has also occurred at many locations.

* registration of all waterwells (including illegal ones) with some 66,000 concessions approved
* introduction of groundwater abstraction charges through a 'water rights fee', although charges remain low and inconsistent (industry US $ 0.073-0.930/$m^3$, municipal up to US$ 0.001/$m^3$)
* transfer of state-drilled waterwells and distribution systems to community/ private ownership/operation
* establishment of 'water rights markets' with the CNA holding a list of selling offers.

• However, the continued exemption of the agricultural sector from abstraction charges is highly anomalous, particularly because some larger-scale irrigators can write-off the capital costs of waterwell construction (US $ 30,000 for 30 l/s yield) within 5 years, given adequate water for double cropping.

| INSTITUTIONAL ARRANGEMENTS | EFFECT ON GROUNDWATER RESOURCE MANAGEMENT | | |
|---|---|---|---|
| | negative | neutral | positive |
| Formal Water Regulation | LACK OF SANCTIONS & ENFORCEMENT | EXISTENCE OF WATER REGIONS<br>CURRENT WATER PRICING POLICY | ADEQUATE WATER RIGHTS DEFINITION & LEGISLATION<br>FORMATION OF AQUIFER MANAGEMENT COMMITTEES |
| Informal Social Norms | CULTURAL NON-COMPLIANCE MENTALITY | | WATER RIGHTS MARKETS & TRANSFERS |
| In Other Sectors | SUBSIDISED ELECTRICAL ENERGY TARIFFS | | REMOVAL OF AGRICULTURAL SUBSIDIES (NAFTA) |

***Inset :*** *Key factors in groundwater resources management*

• The aquifers are mainly either of the:
* intermontane valley-fill type, comprising sequences of alluvial, coalluvial, lacustrine and volcanic deposits which are susceptible to settlement resulting in land subsidence and associated infrastructure damage if subjected to major lowering of the groundwater piezometric surface
* coastal alluvial type, susceptible to saline intrusion if the natural seaward hydraulic gradient is reversed by groundwater abstraction.

• The Mexican Water Law (1992) and subsequent regulations, give extensive powers to the CNA to tackle aquifer overexploitation. Various steps in the process taken during 1993-98 have included (World Bank, 1998):

• A comprehensive study of the factors entering into groundwater resource overexploitation in general, and the incentives and disincentives for individual stakeholders in particular has been undertaken (Inset). This, together with hydrogeologic and socioeconomic modelling of various management scenarios, has led to the following proposals (World Bank, 1998):
* strengthening of aquifer management committees, through financing mechanisms, capacity building and transferral of CNA functions
* building public awareness of the groundwater resource situation to build a consensus for action
* improving groundwater monitoring networks (selective abstraction metering, aquifer piezometric levels, water-use patterns) to provide more useful data.

### KEY ISSUES:

- **addressing the need for progressive reduction of water rights in many 'over-allocated aquifers', including the need for financial support of water-saving technology**
- **re-targetting the electrical energy subsidy to eliminate any incentive for aquifer overexploitation**
- **imposing more realistic municipal water-pricing to provide incentive for reduction in system leakage losses, constraining consumer demand, increasing wastewater reuse and aquifer artificial recharge**

## BOX 10 : Jordan mounts a Primarily Regulatory Offensive to Rationalise Aquifer Exploitation in Extremely Water-Scarce Region

• Jordan provides an important example of an arid country, with low unreliable rainfall (50-600 mm/a) and evidence of extensive overdraft of its aquifers (Inset I), attempting to get to grips with the management of its limited groundwater resources.

• The country possesses 12 aquifer basins but rapid growth in well drilling in the 1980s has led to a total abstraction rate of 465 $Mm^3/a$. Many of the aquifers are being heavily overdrawn with groundwater levels falling at 1-4 m/a and salinity rising steadily; additionally many of the oases have been largely lost (World Bank, 1997).

| GROUNDWATER BASIN | ESTIMATED SAFE-YIELD ($Mm^3/a$) | ACTUAL ABSTRACTION ($Mm^3/a$) 1989 | 1993 | 1997 |
|---|---|---|---|---|
| Amman-Zarqua | 87 | 156 | 184 | 137 |
| Dead Sea | 57 | 82 | 92 | 86 |
| Yarmouk | 40 | 72 | 67 | 52 |
| Al Azraq | 24 | 37 | 50 | 54 |
| Jordan Valley | 21 | 40 | 42 | 36 |

***Inset I:*** *Trends in estimated groundwater abstraction from selected Jordan aquifer*

• Almost 80% of its population of 4.4 million are urban dwellers, but agriculture uses 70% of the groundwater resources to generate about 25% of national productivity and employment. Urban householders have on average spent more than US$500 each on roof and ground storage tanks because of intermittent supply during 6 months/annum. This represents a huge, and viewed globally unproductive, investment. Economic analysis suggests that aquifer storage depletion is best reserved for future urban use (and not for agriculture), since willingness to pay is rising rapidly with higher incomes and population pressures (Schiffler, 1998).

• In the northeastern desert, development of the Al Azraq aquifer is illustrative of the groundwater resources problems experienced. The aquifer has been exploited at levels of 20-25 $Mm^3/a$ for Amman urban water-supply and 30-35 $Mm^3/a$ for irrigated agriculture. However, recharge through intermittent wadi infiltration is difficult to estimate or to verify, because of the thick vadose zone. It is now put in the range 15-35 $Mm^3/a$,even after eliminating most flow to the important Al Azraq oasis.

• In 1995 government embarked on a comprehensive water-sector review and subsequent action plan to confront the water resources crisis, which is amongst the most severe in the world (World Bank, 1997). It was decided that both demand and supply side management measures were urgently needed and in respect of groundwater the following have begun to be implemented:

* agricultural sector investments have been targeted on improving irrigation water-use efficiency, and effecting real water savings through drip technology, and not on extension of irrigated lands
* detailed groundwater basin studies have been undertaken as a precursor to defining management criteria and establishing 'basin protection units'
* by 1997 some 2050 waterwells had been registered and around 1100 abstraction permits issued
* much more severe constraints were imposed in relation to issuing of permits for new waterwells and the replacement or modification of existing ones (Inset II)
* by 1999 meters had been installed at about 75% coverage on operating wells, but resistance is being encountered from a significant minority of private groundwater users, because of fear of escalating water prices or severe constraints on future abstraction
* there has been a public campaign of denouncement of illegal well operators.

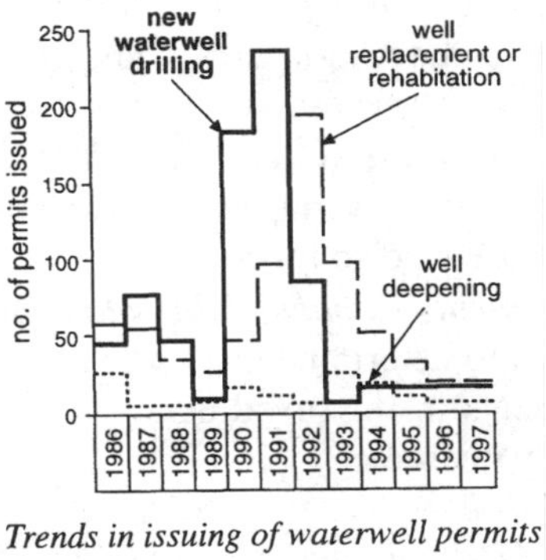


***Inset II:*** *Trends in issuing of waterwell permits in Jordan*

### KEY ISSUES

- **political difficulty has been experienced in relation to groundwater abstraction charges for agricultural irrigation although charges of US$ 0.35/$m^3$ for industrial abstractors and the concept of a levy of US $0.35/$m^3$ on irrigation well owners who exceed their licensed abstraction have been introduced**
- **resistance has also been encountered to reducing agricultural abstraction from the entirely fossil Qa Disi Aquifer, and to reserving its storage for high-value urban and industrial uses**

- ***Reducing subsidies supporting abstraction.*** Many countries provide subsidies for well drilling, pumping equipment and agricultural power, and reduction or removal of these subsidies can help slow the growth in abstraction.

In addition to the above, more intrusive measures such as imposition of well spacing and capacity regulations, water extraction charges, well registration/licensing, restrictions on crop type and/or irrigation technique, and other regulatory, economic and institutional interventions could be used to limit the growth in abstraction. These more intensive management interventions are, however, often only feasible in the context of a full management program.

*Laying the Groundwork for Management*

Once policy measures are in place to limit abstraction, more comprehensive approaches aimed at reducing abstraction can be initiated. It is important to use this phase to build the foundations required for more comprehensive and integrated management.

It would typically include the following types of activities:

- ***Expand public information campaigns.*** These are required to strengthen social support for the implementation of management initiatives, which can be a slow process. Irrigators, in particular, will need to be deeply convinced on the need for reducing their pumping, especially where it is supporting highly-profitable agricultural production. The dissemination of scientific information needs to be clear, transparent, frank about uncertainties and not over-alarmist, otherwise its credibility will be jeopardized.
- ***Establishment of stakeholder involvement.*** If stakeholders are to own management approaches, their involvement is needed before any major decisions are taken.
- ***Development of strategic plan.*** Strategic planning consists of a systematic review of available information to identify broad areas where action is needed, of the identity of key stakeholders and of the approach that will be taken to addressing overdraft concerns. It should *link the sets of actions required for management with the institutions that will be responsible for implementing them.*
- ***Establishment of legal basis for management.*** This involves legislation enabling the declaration of resource conservation zones and the formation of regulatory agencies and aquifer management committees, as necessary.
- ***Well registration system.*** Establishment of a system for registering all wells is important as the foundation for any subsequent activities to evaluate and monitor abstraction, establish water rights, regulate use, and so forth.
- ***Improvements in basic data collection.*** In many countries groundwater data collection systems are sufficient to determine that overdraft may be occurring but insufficient to actually identify solutions. Investments in improved data collection targeted on particular regions of concern are important in this early phase.

*Beginning to Reduce Abstraction*

Once a solid basis for resource administration exists, initiatives to reduce abstraction on a relatively large scale can, if necessary, be taken. Because the results of strategic planning will reflect local needs and conditions, approaches will diverge increasingly in different contexts. As a result, the elements indicated are likely to differ greatly in detail between locations.

Overall, however, they would typically include:

- ***Expansion of information dissemination.*** Many of the activities taking place during this first stage of management implementation are likely to be socially and politically sensitive. As a result, education and information will be important in order to maintain a broad base of social support.

- ***Consolidation of regulatory system.*** Potential regulatory approaches beyond the initial ones designed to slow the growth in abstraction should have been identified. The most common approach is to require permits for new well construction above a certain capacity or depth, and place limitations on well spacing, drilling depth, well diameter and pump capacity.
- ***Initiation of water rights reform.*** The strategic planning exercise will probably have identified a large set of water rights and allocation issues. Resolution of these issues would typically be a major activity during this phase and would consist, for example, of participatory processes to reform water rights and of legal measures to establish regulatory bodies and/or water markets.
- ***Improvement of water-use efficiency.*** Promulgation of efficient water-use technologies and other approaches to reducing water consumption (such as low water-use crops) would be a major activity. It would involve the establishment of extension capabilities, demonstration activities and collaborative research to develop and disseminate technologies. It will be important to focus primarily on those changes which are likely to maximize "real-water savings".
- ***Establishment of resource conservation zones.*** This consists of identifying zones for management and setting up the organizations such as aquifer management committees that will implement management activities within them. In most cases, this would be done initially on a pilot basis in areas where *both* management needs *and* options are clear. Governments often want to address the most problematic areas first, but this can be a recipe for failure because problems are so entrenched that politically and socially viable reductions in abstraction will have little impact. Strategically, it may be better to focus on areas where problems are not too advanced, technically viable options are readily available, and there is strong social support for management.
- ***Implementation of economic incentives for conservation.*** A variety of economic incentives for water conservation will have been identified in the strategic planning process. These might consist, for example, of water abstraction charges, pump taxes, energy price increases, subsidies for water conservation equipment or imposition of taxes on water intensive crops and can be implemented locally or at the macro state level.

*Scope for Artificial Aquifer Recharge*

The possibilities of artificial aquifer recharge to support groundwater resource development in the rural context will range from:

- Individual small-scale measures designed to enhance the infiltration of wet season run-off either at field level or in the beds of small watercourses
- Formal artificial recharge schemes where excess surface water flows are directed to infiltration basins.

The option of aquifer recharge via boreholes is not considered economic in the rural context, because of the high level and cost of operational treatment required.

There is little doubt that improved soil tillage and terracing, together with modified cultivation regimes at field level can much reduce soil erosion and enhance groundwater recharge. In some hydrogeological conditions (especially permeable hill country and on alluvial outwash fans), small check dams constructed on surface watercourses can provide an increased rate of riverbed recharge, although there is little data to prove the efficiency of this method and some question about sustainability due to silting-up. In view of their low capital cost, however, such measures should generally be encouraged, using detailed local hydrogeological knowledge to improve their siting and design.

The scope for larger-scale artificial aquifer recharge schemes using infiltration basins will be determined by a number of factors:

- The vertical permeability of the subsoil and the existence of sufficient infiltration capacity to depth to permit significant rates of groundwater recharge

- The availability of land for the construction of infiltration basins in areas with suitable hydrogeological conditions, and in proximity to the proposed source of excess surface water (this land requirement will include the need for pre-recharge sedimentation basins of appropriate dimensions)
- The availability (volume and duration) of excess surface water of acceptable quality for aquifer recharge
- An institutional and organizational structure capable of promoting and operating such a scheme, bearing in mind that it is not always easy to recover costs directly from groundwater users, and these (at best) would have to be levied via groundwater abstraction permits.

The design of artificial aquifer recharge schemes is not straightforward and requires significant hydrogeological site investigation, followed by a pilot-scale operation with detailed monitoring and performance appraisal. The factors described above limit considerably the geographical area potentially suitable for such schemes, but more emphasis nevertheless needs to be put on their investigation and promotion, especially in areas where there already exists a considerable overdraft on aquifer storage.

*Option of Planned Mining of Groundwater Storage*

It should be pointed out that there is no fundamental reason why the overdraft of aquifer storage is an undesirable process. If the practice of mining groundwater reserves is carried out on a carefully planned basis, it can form part of a logical water resources management strategy (Foster, 1992; Lloyd, 1997).

For this to be the case, however, the groundwater system under consideration should be sufficiently well investigated and understood to evaluate reliably the following:

- The rate of groundwater mining that can be achieved for the period in question
- The scale of any internal effects on the aquifer system and of any external impacts on the environment
- The level of interference with all existing, and potential future, users of the groundwater resource
- An economic analysis of the benefits of groundwater mining for the proposed use, compared to those of alternative and future uses.

It is strongly recommended that a systematic evaluation of these criteria be undertaken as part of a strategic analysis of water resource management options, before a conscious decision to mine groundwater storage is made. This will normally require a sizeable program of associated hydrogeological investigation.

All too often, however, this is not the case and a sequence of progressive overdraft of aquifer storage is embarked upon in an anarchical or unplanned fashion, with negative long-term consequences for all groundwater users. This is more especially the case in aquifers in which some limited current recharge is occurring than in the case of aquifers containing essentially "fossil" storage.

# 4

# *PROTECTING GROUNDWATER QUALITY IN RURAL AREAS*

## Nature of Diffuse Pollution Threat from Agriculture

Over the past few decades there has been a radical evolution in agronomic practice in many regions of the world associated with (largely successful) attempts to increase agricultural productivity. The intensification of production from agricultural land has been sustained by the application of ever-increasing quantities of inorganic fertilizers and a wide spectrum of synthetic pesticides. In the more arid regions, cropping frequency has been increased and additional land has been brought into production through new irrigation schemes and increasing irrigation efficiency. A common trend is the replacement of traditional crop rotations by intensive and continuous cultivation of high-value crops, selected according to prevailing market conditions. In many instances, near monocultures across extensive tracts of agricultural land have resulted.

In many nations the principal recharge areas of lowland aquifers form valuable tracts of farming land and are now almost completely used for *intensive crop cultivation.* In such cases the bulk of replenishable groundwater resources originate as excess rainfall and excess irrigation infiltrating this land (Figure 2). As a consequence, these resources are vulnerable to contamination by cultivation practices. The large extent of agricultural activities makes the impact all the more significant.

There is a risk of elevated rates of *nutrient, salt and pesticide leaching to groundwater* from cultivated soils with the corresponding potable water-quality guidelines (Table 16) being exceeded. This is especially the case in areas of well-drained (thin and/or sandy) soils widely found in aquifer recharge areas.

**Table 16: Summary of water-quality guidelines related to groundwater contamination through agricultural cultivation**

| *Water quality guideline* | *Parameter concentration* | | | | | |
|---|---|---|---|---|---|---|
| | *$NO_3$ (mg/ℓ)* | *Cl (mg/ℓ)* | *Na (mg/ℓ)* | *$SO_4$ (mg/ℓ)* | *Insecticides (m (μg/ℓ)* | *Herbicides (μg/ℓ)* |
| WHO (potable) | 45 | 250 | 200 | 400 | 0.1-30** | 2-100** |
| US-EPA (potable) | 45 | 250 | - | 250 | 0.1-10 0** | 30-100** |
| EC (potable) | 50* | 200 | 150 | 250 | 0.1 | 0.1 |
| US-ARS (irrigation) | - | 100** | 100** | - | - | 100** |

* EC also give maximum concentrations for $NO_2$ (0.1 mg/ℓ) and $NH_4$ (0.5 mg/).

** for most sensitive crops, many others can tolerate 5-10 times higher concentrations.

*Note:* The range for individual listed insecticides or herbicides is given, but many remain to be evaluated because of lack of medical evidence; Se and As, which may also be leached from some irrigated soils, are not included.

Prior to the late 1970s there was widespread complacency about such risks. Environmental regulators and agricultural administrators have been slow to recognize the scale of potential problems, due to a number of contributory causes:

- Generally slow average rates of vadose zone transport of contaminants leached from cultivated soils (which even for non-reactive pollutants do not normally exceed 2 m/a and 5 m/a beneath non-irrigated and irrigated land respectively) and the resultant delayed impact (or legacy) for groundwater quality
- Lack of consciousness among groundwater specialists about the level of influence of changes in agricultural land-use practices, due in part to limitations in sampling from the vadose zone
- Preoccupation in the agricultural sector with problems reducing agricultural productivity (such as waterlogging and soil salinity due to rising groundwater table), but not with groundwater quality deterioration.

## Processes Controlling Nitrate Leaching and Transport

Agricultural soils contain large, but widely varying, quantities of nitrogen in organic form, often amounting to more than 2000 kgN/ha/a. This is oxidized by soil bacteria to soluble nitrate (at rates varying with soil temperature and humidity), which is then susceptible to leaching below the root zone. *Inorganic N fertilizers* are added to increase the immediate availability of nitrate for plant growth, while manures (which also contain large quantities of less readily available nitrogen) are applied primarily to replenish soil organic matter.

The nitrogen in plant nutrients applied to the land is subject to complex soil processes. It may be taken up directly by the growing crop, incorporated into the soil N pool, reduced and lost in volatile from (as $NH_3$ or $N_2$ gas) or as nitrate by soil leaching or in surface run-off. Thus while only a small proportion of the nitrate leached in a given year is derived directly from inorganic fertilizers, the overall rate of nitrogen mineralization and leaching normally relates in a general way to fertilizer application rates.

The *leaching of nitrate* from dryland agricultural soils is dependent on a complex interaction of soil type, cropping regime and rainfall infiltration, which cause significant uncertainty when estimating the average rates of loss (Vrba and Romijn, 1986; Foster, 1989; Spalding and Exner, 1993). Some leaching from the soil will occur when no nitrogen is applied and/or the land is fallow. In some arid climates the concentrations of nitrate leached from beneath natural vegetation are also high.

In-situ natural *denitrification in aquifer systems* has been the subject of considerable research (Korom, 1992), because it results in removal of nitrate from groundwater. If active on a widespread basis, it can have a major beneficial effect on groundwater quality. Clear evidence of denitrification comes from some confined aquifers (Lawrence and Foster, 1986). The process is likely to be bacteriologically mediated, and clear evidence of appropriate bacteria has been found; but it could also be chemical, accompanying the oxidation of disseminated pyrite found in many geological formations.

In the vadose zone, the generally aerobic conditions and persistence of high nitrate concentrations to depth imply that denitrification cannot be widely active, despite the presence of potentially-denitrifying bacteria, but it may be more significant in the zone of water table fluctuation. However, where the unsaturated zone includes strata rich in organic carbon, the process may become more predominant.

*Sources of nitrate* can be distinguished by the analysis of nitrogen isotopes and of associated elements (such as Cl) in groundwater. Naturally-mineralized soil organic nitrogen, inorganic fertilizers, human/animal wastes and precipitation have distinctive but overlapping isotopic ($d^{15}N$) signatures, and this has been put to use with varying success in a number of studies. Groundwater nitrate originating from inorganic fertilizers is clearly distinguished from that deriving from organic sources, but it is not possible to distinguish between human and animal sources. Moreover, the isotopic signatures can be modified by ammonia volatilization and denitrification, although by using both the $^{15}N$ and $^{18}0$ isotopic signatures of $N0_3$ it is possible to investigate both the origin and fate of nitrate in groundwater systems.

Where irrigation is practiced, there exists the possibility of controlling soil moisture so as to maximize nutrient uptake and to restrict deep percolation, thereby controlling the leaching of agrochemicals. This is most practicable where virtually all plant moisture requirements are provided by irrigation and where the wet season is confined within a few months each year. It is less feasible where irrigation is required mainly to secure a second crop, but even here the maximization of nitrate uptake can be assured by providing optimum moisture levels at times of rapid plant growth, and thereby reduce soil nutrient residues. Moreover, denitrification losses become more significant in irrigated cultivation, at least on finer-grained soils. However, many irrigation practices remain relatively inefficient with excess moisture applied by each irrigation lamina. On freely-draining soils especially, regular soil leaching and deep percolation of nitrate results, and this has been judged the cause of steadily-increasing concentrations of groundwater nitrate in many irrigated areas (Foster, 1989; Foster and Chilton, 1998).

In northwestern Sri Lanka, for example, intensive irrigated horticulture is being carried out on permeable sandy soils over a shallower calcareous sand aquifer. Triple cropping with applications of up to 500 kgN/ha/a is producing significant nutrient losses (Figure 3), with high nitrate and occasional ammonium in groundwater. A *close correlation is observed between land-use and nitrate concentrations* in the underlying groundwater. More detailed research in this area suggested that equivalent to 70 percent of the fertilizer N application was being lost from agricultural soils, after taking account of the recycling of nitrate in irrigation water.

Although there has been relatively little detailed investigation of nutrient leaching to groundwater under cultivation practices typical of humid tropical regimes, a more favorable picture emerges from some extensive irrigated cultivation on less permeable soils (Box 11). Since *pasture is less prone than cultivated land to nitrate leaching*, it offers a useful option for controlling aquifer nitrate pollution. However, detailed investigation has demonstrated a constraint in that leaching rates from pasture on well-drained soils increase abruptly and unpredictably to elevated levels when grassland productivity is intensified by heavy applications of nitrogen fertilizer and by high density grazing.

Where *highly efficient irrigation techniques* have been introduced, greater control over leaching is possible, but groundwater recharge will become progressively more saline (Foster, 1989). Nutrient leaching will be much reduced, but nitrate concentrations may remain high because of the much smaller volume of deep percolation. While controlled precipitation of $CaCO_3$ and $CaSO_4$ in the vadose zone is theoretically possible, field evidence suggests that with increasing irrigation efficiency, *salinization of groundwater recharge* can cause severe problems in arid regions, especially those where groundwater itself is the only source of irrigation water and recirculation with progressive fractionation occurs.

This is not the only way in which irrigated agriculture can cause groundwater salinization (Chilton, 1995). Other processes include:

- Inefficient irrigation with imported surface water over semiconfined aquifers with relatively low-permeability surface strata, which can lead to rising water table, soil water-logging, phreatic evapotranspiration and, in consequence, *salinization of soils* and shallow groundwater, where soils become water saturated with sodium in excess of 60 percent of the cation exchange capacity, their alkalinity can rise to pH 9-11 and degradation by compaction often results
- *Leaching of natural salts from desert soils* (and the vadose zone) by infiltrating excess irrigation, this has occurred in some parts of the Yaqui Valley in Sonora-Mexico (with post-irrigation recharge containing more than 1000 mg/ℓ) and in the Murray River basin of South Australia (but in this case increased infiltration was caused simply by clearing of natural semi-arid vegetation).

## Risk of Pesticide Contamination

All pesticide compounds pose a significant environmental health hazard since they are, to greater or lesser degree, chemically tailored to be toxic and persistent. However, prior to 1980, there was not much concern about the possibility of groundwater pollution by pesticides, since agricultural scientists argued that soil sorption of the higher molecular weight compounds (such as the chlorinated hydrocarbon insecticides) and volatilization of lower molecular weight compounds (like most herbicides) would predominate.

## *BOX 11 : Leaching of Nitrate from Tropical Agricultural Soils to Groundwater*

• Data on nitrate leaching from tropical agricultural soils is limited, and much of the published data (Foster & Chilton, 1998) is summarised here.

• In the major Mexican semi-arid wheat-growing area of the Yaqui Valley-Sonora, current fertiliser application rates range according to crop from 120-220 kgN/h/ha, (in the form of urea and anhydrous ammonia). Usage increased steadily from the 1960s becoming constant during the 1980s. Profiles obtained from deep investigation boreholes, together with regular sampling of selected shallow irrigation wells, shows nitrate concentrations in groundwater recharge in the range 10-25 mg/l (Chilton et al, 1995), tolerable from the point-of-view of drinking water provision.

• In the humid tropics, where crops often require supplementary dry-season irrigation, it is believed that greater moisture availability and higher soil temperatures result in good N uptake by plants and modest nitrate leaching, at least by traditional crops. High clay mineral and organic matter content in deeply-weathered tropical soil profiles may also favour denitrification.

• Barbados has a long history of sugarcane cultivation. Sugarcane receives about 550 kg/ha/y of 24N-OP-18K fertiliser, amounting to about 130 kg N/ha/a. Some of this may be subject to direct leaching when it is applied. However, sugarcane is overall an efficient user of nutrients because of the continuous crop-cover with strong root development. Currently, nitrate concentrations in most wells in the highly-vulnerable limestone aquifer are also in the tolerable range of 25-35 mg/l, which is consistent with leaching losses of 40-60 kg N/ha/a (Chilton et al, 1995).

• In Queensland-Australia, the fate of N fertilisers applied to sugarcane and pasture land in an area with a high mean rainfall of 3200 mm/a and freely draining soils (with mainly rainfall infiltration of 710-1260 mm/a) has been investigated (Prove et al, 1994). Application rates to sugarcane are 160-180 kg N/ha/a and to pasture land from 0-500 kg N/ha/a in 100 kg/ha/a splits. Leaching losses in the same period averaged 60 kg/N/ha for sugarcane but were insignificant under pastureland. At all sites most of the nitrogen leached moved as nitrate in a rapid pulse following heavy rainfall, but resultant average concentrations are not excessive due to dilution from very high infiltration rates.

• In view of its very widespread distribution in southern and eastern Asia, paddy cultivation warrants special consideration and a layered alluvial aquifer in the Madras area of India has been studied (Chilton et al, 1995; Foster & Lawrence, 1995). Typical annual cultivation cycles consist of two rice and one groundnut crop, each receiving at least 60 kg N/ha/a. Monitoring of groundwater quality in piezometers constructed in the upper aquifer immediately beneath rice fields enabled the quality of the recharge from cultivated soils to be assessed and demonstrated low nitrate concentrations (Inset). One explanation for the low concentrations is that denitrification is active for part of the year in these anaerobic flooded soils, although this may not be the case for paddy cultivation under all soil conditions.

***Inset :*** *Groundwater nitrate concentrations below paddy cultivation near Madras, India*

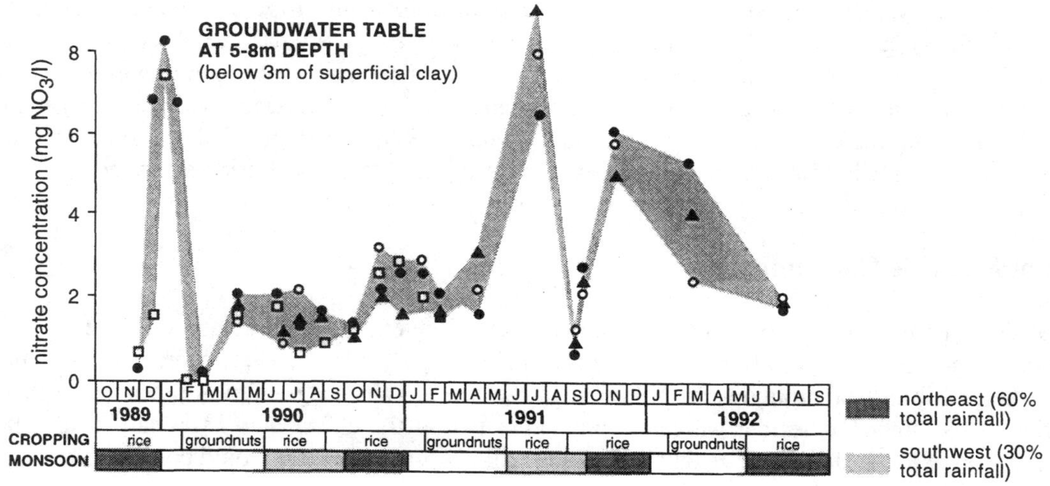

Investigations to appraise adequately the level of *pesticide leaching* to groundwater require analysis of samples from the water table in aquifer recharge areas, and this is proving costly and problematic because of:

- The wide range of pesticide compounds in common use, many of which also break down to *toxic derivatives* (metabolites)
- The need to work at very low concentrations because of the high toxicity of many compounds, which necessitates the collection of large sample volumes and careful handling to avoid compound modification and volatile loss.

In view of these difficulties, an essential prerequisite is to identify the most likely types and sources of pesticide contamination and the most probable mechanisms of transport from the land surface to groundwater. Such information is essential for the specification of sampling protocols and monitoring networks, and to prioritize and rationalize investigation work (Box 12).

Rates of agricultural pesticide application are generally in the range 0.2-10.0 kg/ha/a of active ingredient. Many pesticide compounds have water solubilities in excess of 10 mg/ℓ and this is not a limiting factor in leaching from soils. Of greatest importance in this respect is their *mobility in soil solution*. This will vary with affinity for organic matter and/or clay minerals, and can be expressed by the corresponding partition coefficient, which is normally available in manufacturer listings, but only for adsorption onto soil organic matter. An important anomaly in respect of subsurface pesticide mobility is the fact that some, otherwise strongly-adsorbed, compounds could be mobile in fissured or coarse-grained formations in the sorbed phase, if attached to colloidal particles.

Both the *mode of application and action of the pesticide* are important factors in relation to soil leaching, since those targeted at plant roots and soil insects are much more mobile than those acting directly on plant vegetation (Foster and Chilton, 1998). Chemical reactivity of the compound with the soil matrix may also play an important role in reducing the risk of pesticide leaching, as a result of the generation of less soluble residues, for example, through neutralization of acidic compounds in alkaline soils.

The *degradability of pesticide compounds* in the soil horizon, as a result of bacteriological oxidation or chemical hydrolysis, will normally be significant. Soil half-lives for most compounds currently in widespread use range from 10 days to up to 10 years, but for the more mobile compounds are normally less than 100 days. However, given the timing of applications, they are sufficiently persistent to remain in the soil for significant periods when leaching may occur. Moreover, some derivatives of partial oxidation or hydrolysis (metabolites) may be equally toxic and/or mobile as the original pesticide compounds themselves (Kolpin and Goolsby, 1995).

Pesticide compounds leached from permeable soils into the vadose zone enter an environment which contains a much smaller proportion of clay minerals and organic matter, and has a greatly reduced population of indigenous bacteria. Thus the mobility and persistence of all pesticide compounds should be expected to be many times greater in the vadose zone than in a typical agricultural soil.

Nonetheless, under the vast majority of conditions it is unlikely that downward matrix transport rates in the vadose zone will exceed 1 m/a and for many compounds they would be very much slower than this value. Since few pesticide compounds have been in regular widespread use for more than 10-20 years, most pesticides leached from agricultural soils would be expected to have been degraded or still to remain in the vadose zone, if this was the only transport mechanism operative, except in areas of shallow groundwater table.

## *BOX 12 : Risk of Pesticide Leaching from Tropical Agricultural Soils*

• The occurrence of pesticide residues in the vulnerable limestone aquifer of Barbados, where the herbicides atrazine and ametryn are applied widely to sugarcane at rates of around 4 kg (ai)ha/a, has been investigated (Wood & Chilton, 1995). Atrazine, and its derivative (metabolite) deethylated-atrazine, were regularly detected in groundwater at concentrations in the range of 0.5-3.0 μg/l and 0.2-2.0 μg/l respectively (Inset I).

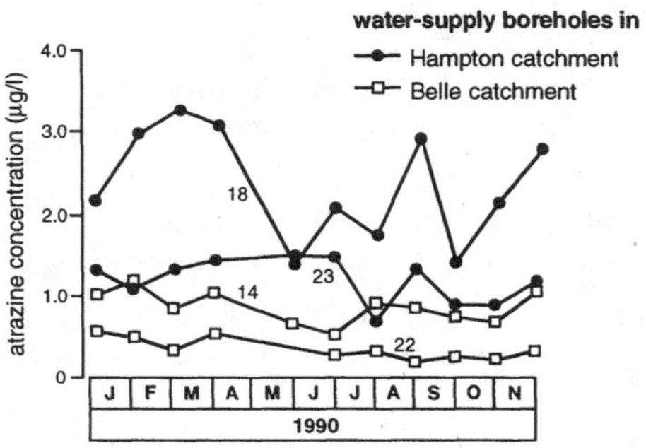


*Inset I: Groundwater atrazine concentrations in Barbados catchments under sugarcane cultivation*

• Research has also been undertaken on the northwestern coast of Sri Lanka on the fate of carbofuran (Foster & Lawrence, 1995), which was applied at 6 kg (ai)/ha to horicultural crops. The parent compound is highly mobile and was rapidly leached from the soil with concentrations of 200-2000 μg/l in the soil drainage of a lysimeter and peak concentrations in excess of 50 μg/l in the underlying shallow groundwater within 20 days of application (Inset II). Carbofuran was, however, subject to rapid degradation and in part transformed to its more persistent (but less mobile) metabolite, carbofuran-phenol. This remained in the shallow groundwater for more than 50 days.

• Although available research and monitoring is very sparse, there is sufficient to demonstrate the risk of leaching of agricultural pesticide to shallow groundwater in highly vulnerable aquifers, and the potential persistence of toxic compounds in these systems. It is, however, not possible to make a realistic assessment of the risk of contamination of deeper groundwater in less vulnerable aquifers (Foster & Chilton, 1998).

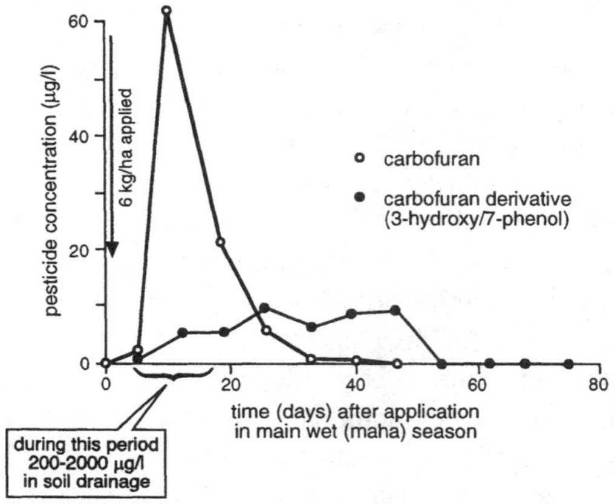


*Inset II: Leaching of the insecticide Carbofuran from irrigated cultivation to shallow groundwater in vulnerable aquifer on Sri Lankan coast*

• Given the wide range of pesticide compounds in use in agriculture, and their many toxic metabolites, an approach to groundwater pollution risk assessment based on the key properties of the pesticide compounds (mobility, solubility) and of the geological media (propensity to preferential flow in vadose zone) is needed to target monitoring.

• In general terms, a significant additional element of protection for drinking water-supplies will be provided if their intake is at a considerable depth below the water-table, and the sanitary integrity of upper section of solid well casing is sound. This will generally provide additional aquifer residence time for pesticide degradation before entry to the waterwell concerned. Those wells most vulnerable to contamination by agricultural pesticides will be shallow dug wells providing domestic supplies to isolated rural farmsteads in areas of intensive cultivation.

The hydraulic characteristics of many aquifers are, however, such as to present high probability of the development of so-called *preferential flow in the vadose zone*, especially (although far from exclusively) in consolidated fractured formations. Preferential flow is of major importance in the consideration of pesticide transport into aquifers (Foster and Chilton, 1998). Where developed, it would provide routes for deeper penetration of readily-leached pesticide compounds and would be characterized by much more rapid pollutant transport, providing less opportunity for retardation through molecular diffusion into the microporous matrix and associated adsorption, chemical reaction, and biodegradation. If preferential flow in fissures of larger aperture occurred, the possibility of transport of less mobile pesticide compounds adsorbed on colloidal material would also arise.

There has, as yet, been very little research and monitoring of the leaching of agricultural pesticide residues and derivatives under tropical conditions, but some limited available data are given in Box 13.

## Controlling the Leaching of Agrochemicals

The preceding sections demonstrate that agricultural cultivation can have a significant impact on groundwater quality and, under some conditions, seriously compromise its value as a primary source of potable water supply. In a qualitative sense Table 17 indicates the relative influence of hydrogeologic and agronomic factors in this process. It also indicates in a general way (through bold type) those factors which to some degree can be controlled by changing cultivation type or practice.

In more general terms a rational strategy for the control of diffuse groundwater pollution from agricultural cultivation practices would include the following measures (Foster and Chilton, 1998):

- Recognize that incremental changes in the intensification of agricultural cultivation can run high risk of adverse impact on groundwater quality, while offering rather marginal returns to farmers
- Adopt major aquifer recharge areas as a separate unit in guidelines for agronomic practice, taking account of the need to reduce leaching to groundwater
- Introduce groundwater leaching assessment in cropping trials before new agronomic practices are recommended and pesticide compounds approved
- Accept that more positive control over land use may have to be taken in groundwater source protection areas.

**Table 17: Summary of the relative impact of agronomic factors on groundwater quality**

| SOIL LEACHATE CONCENTRATION* | | DETERMINING FACTORS<br>RANGE<br>lesser ------------------ greater | SOIL LEACHATE CONCENTRATION* | |
|---|---|---|---|---|
| **Nitrate** | **Pesticides** | | **Nitrate** | **Pesticides** |
| - | - | soil permeability | + | + |
| ++ | ++ | soil thickness | - | -- |
| + | - | excess rainfall | - | + |
| ++ | ++ | **irrigation efficiency**** | - | -- |
| ++ | + | **control of applications** | - | -- |
| 0 | -- | **pesticide (type) mobility** | 0 | ++ |
| ++ | - | **continuity of cultivation** | -- | + |
| - | + | **frequency of plowing** | ++ | - |
| -- | 0 | **grazing intensity**** | ++ | 0 |

*Note:* It is difficult to be more prescriptive than this due to the wide range of agricultural regimes and hydrogeological conditions under potential consideration.

* concentration not load since latter also requires consideration of recharge volume
** where applicable
(+)+ tends to increase concentration
0 minimal effect
(-)- tends to decrease concentration

*Source:* Foster and Hirata, 1988.

## *BOX 13 : Groundwater Source Pollution Risk Evaluation & Management around Managua, Nicaragua*

• Groundwater is of the utmost importance for domestic, industrial and agricultural water-supply in the region around Managua, which has a population well in excess of 1.0 million. Water is extracted from deep municipal and private boreholes in the major volcanic aquifer system located south of Lake Managua, (deposited by eruptions of the Masaya Volcano, whose crater is situated some 20 km southeast of the city).

• The volcanic formations include lava flows from the volcano (last major eruption 1792), interbedded with pyroclastic deposits. There is little soil development on the most recent flows and no surface run-off with high rates of rainfall infiltration/ groundwater recharge. The area is classified as highly-vulnerable, despite the relatively deep water-table (ranging from 25 m bgl to more than 100 m bgl close to the volcano), except where alluvial-volcanic deposits of lower permeability occur at the surface.

• The main existing wellfield abstracts some 195 Ml/d and is located in the urban fringe east of Managua City, but a new wellfield at a more rural location some 10 km south of the city is under investigation and development.

• The entire area, including the groundwater capture zone of the proposed new wellfield of 70 Ml/d, has been the subject of systematic groundwater resource risk evaluation, including aquifer vulnerability mapping and subsurface contaminant load survey (Scharp, 1994; Scharp et al 1997). In this work there was a clear policy to involve all stakeholders; not only the major users but also the potential polluters of groundwater.

• The capture zone of the existing wellfield is threatened by a range of industries including tanneries, metal workshops and textile manufacturers in the Zona Franca industrial area, as well as fuel and chemical storage at the international airport and a number of developing periurban towns with in-situ sanitation. There are also several small air strips in the area, which were historically used for storage, loading and aerial spraying of agricultural land. In the past 30 years there was intensive cotton cultivation using many highly persistent pesticides, such as Toxaphene and DDT.

• The predicted capture zone of the new wellfield is classified as having mainly moderate vulnerability, but there are areas of high vulnerability due to the absence of soil cover, which has been removed through erosion. While there are a number of potential point sources of contamination from industry, petrol filling stations and waste disposal sites, only one industrial site with underground storage tanks has been classified as having high potential contaminant load.

• The capture area is more predominantly agricultural and it is considered that the frequent use of mobile pesticides (such as the carbamate insecticides) poses the major pollution threat, and control over agricultural activity will be needed in the interests of municipal water-supply.

***Inset :*** *Groundwater pollution assessment mapping for Managua groundwater system (eastern area)*

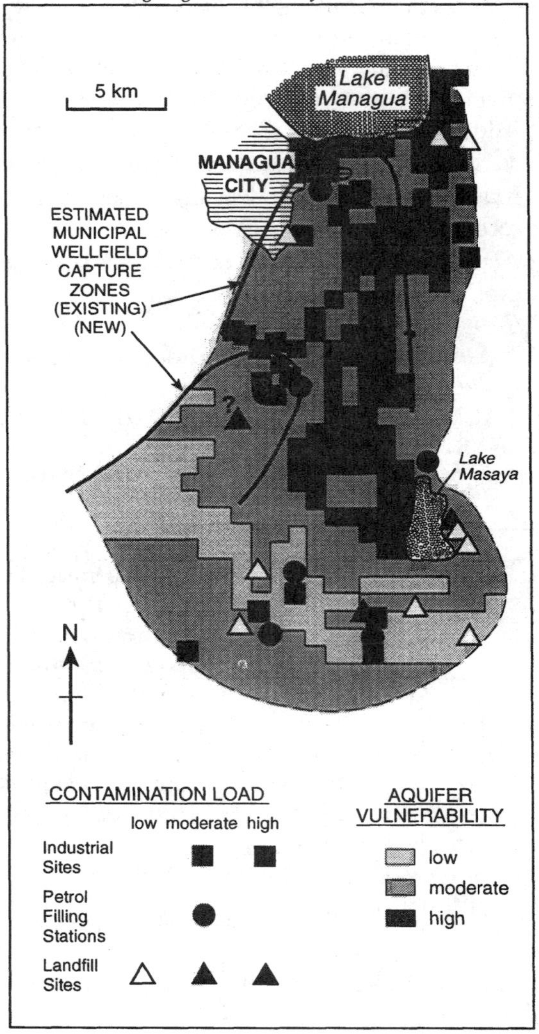

Agricultural pollution stems from literally millions of everyday activities and management decisions made by farmers. Individually these activities may not cause discernible environmental harm, but the aggregation of these activities over many months or years can combine to affect groundwater quality adversely, and even the productivity of the soil itself. Application of a regulatory approach to diffuse pollution has the significant problem of identifying both measurable and enforceable standards and the resources needed to monitor compliances on millions of acres of agricultural land, and is a formidable challenge. There is thus a need to identify priorities both in terms of the more polluting aspects of agricultural practice and the groundwater resources most in need of protection.

Groundwater quality protection creates the need to work with all agricultural producers, and particularly those who are motivated to care for their land. Pollution prevention strategies need to focus on improved management practices leading to contaminant source reduction or risk management. Some government programs intervene when environmental problems present direct threats to health. Such intervention is crisis management and not pollution prevention. While pollution prevention may require some investment in new technologies, more often it is a matter of improving behaviors and practices.

Contaminated groundwater will threaten the farmers who pollute, as well as the community drinking water supply. There are thus incentives for farmers to practice prevention. Of course, farmers alone cannot be expected to meet the challenge of incorporating pollution prevention into agriculture, and governments must provide some incentives and support initiatives.

Agriculture includes not only cropland and pastures, but also farm buildings and facilities. These locations also involve groundwater pollution risk. Farmsteads can have petroleum tanks, pesticide and fertilizer storage units, septic tanks or pit latrines, livestock yards, feedstuff and manure storage facilities. The concentration of potential contaminants and intensity of activities around farmsteads can generate significant pollution risks from nitrates (Figure 20), toxic chemicals and microorganisms, especially to the domestic waterwells on (and in the vicinity of) farms.

## Pollution Hazard Assessment and Protection Strategy

### *General Approach*

Improving the protection of groundwater against serious pollution is a complex task, involving concepts that are not widely understood. Two interrelated but independent components should be recognized, namely protection of:

- Groundwater resources or aquifers as a whole
- Groundwater sources, that is, those parts of aquifers where the resource is exploited for potable water supply.

The latter is normally considered as supplementary to the former, but a realistic balance between the two needs to be struck, according to local circumstances (Foster and Skinner, 1995).

Aquifers are naturally (but variably) protected against pollution of their groundwater by the vadose zone or the confining beds which overlie them. For groundwater protection policy not to be unnecessarily restrictive on human economic activity, this *natural attenuation capacity* must be utilized. This can be achieved by zoning the vulnerability of the underlying aquifer to pollution at the land surface, and thus enabling priorities for pollution control to be logically assigned. Controls would be sought over existing and new activities involving potential hazards to groundwater, according to their location in relation to such zones.

**Figure 20: Groundwater nitrate concentrations in the weathered basement aquifer of rural areas of central Nigeria**

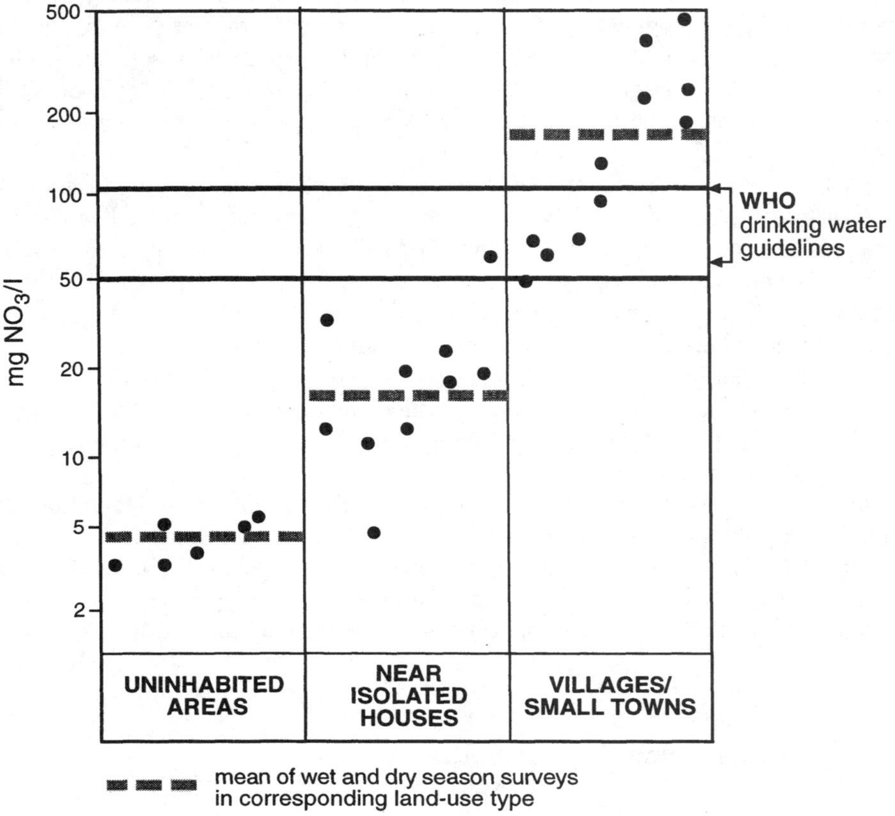


*Note:* The uninhabited areas were in use for low-intensity dryland cropping and animal grazing; the increase in groundwater nitrate concentrations near habitation is due to in-situ disposal of human and animal excreta
*Source:* Langenegger, 1994.

In areas with intensive agricultural development, the zones would serve to define the priority for establishing an *inventory of hazardous chemicals*, for *estimating subsurface contaminant load* due to soil leaching and for designing an *aquifer monitoring network* (Chilton and others, 1990). Such actions would be required before the implementation of pollution control measures could be rationally justified.

The need to achieve maximum aquifer protection will also vary with the utilization (actual or designated) of groundwater resources. Protection measures should normally be intensified around public water supply sources. Thus in the assessment of groundwater pollution hazard and the formulation of groundwater protection policy, the basic prerequisite is both:

- The ranking and mapping of aquifer pollution vulnerability
- The definition of special groundwater source protection areas.

These tasks are discussed further technically in succeeding sections. In socioeconomic terms they are effective vehicles for initiating the involvement of all stakeholders (including water supply interests and potential agricultural polluters), which will be essential if progress on groundwater quality protection is to be made (Box 14).

The emphasis placed on one or other of these approaches will depend on the resource development situation and prevailing hydrogeological conditions (Foster and Skinner, 1995). Source-oriented strategies are best suited to more uniform, unconsolidated, aquifers exploited by a relatively small and fixed number of high-yielding public water supply boreholes with stable pumping regimes. They are particularly appropriate in sparsely-populated regions where their definition can be fairly conservative without producing serious conflict with other interests. They cannot be so readily applied where there are rapidly growing numbers of individual abstractions and seasonally-variable pumping, since this will render consideration of individual sources and the definition of fixed zones impracticable. Data deficiencies and scientific uncertainties, especially in heterogenous aquifers, can also render the estimation of protection zones inadequate.

Aquifer-oriented strategies are more universally applicable, but it has to be recognized that there may be limited parts of aquifers which do not justify protection because their water quality is naturally too poor or has already suffered excessive deterioration. A further complication arises where groundwater systems are thick and layered, and it will be essential from the outset to be clear about which aquifer is being considered.

### *Mapping Aquifer Pollution Vulnerability*

The ability of natural subsoil profiles to attenuate many water pollutants has long been implicitly recognized. To a lesser degree, the attenuation processes continue below the soil, deeper in the vadose zone, especially where unconsolidated sediments, as opposed to consolidated fissured rocks, are present.

However, not all soil profiles and underlying hydrogeological environments are equally effective in pollutant attenuation. Moreover, the degree of attenuation will vary with types of pollutants in any given environment. Concerns about deterioration of groundwater quality relate principally to unconfined or phreatic aquifers, especially where their vadose zone is thin and their water table is shallow. A significant pollution hazard may also be present even if aquifers are semi-confined and the overlying aquitards are relatively thin and/or permeable. Groundwater supplies drawn from deeper, highly confined aquifers are much less affected by pollution from the land surface, except by the most persistent pollutants in the very long term.

*Aquifer pollution vulnerability* is a helpful concept increasingly used to indicate the extent to which an aquifer can be adversely affected by an imposed contaminant load. This is a function of the *intrinsic characteristics of the vadose zone or the confining beds* that separate the saturated aquifer from the land surface immediately above (Foster and Hirata, 1988). Some hydrogeological environments are inherently more vulnerable than others (Table 18). Areas of the same aquifer system may have different relative vulnerability due to spatial variations in vadose zone thickness or the character of confining strata. Mapping of aquifer pollution vulnerability provides a simple, but consistent, set of criteria for *land surface zoning*. The integrated vulnerability concept is not scientifically precise, but the concept provides a general framework within which to base groundwater protection policy and pollution control measures (Table 19).

Where the leaching of agricultural chemicals is the major concern, the scheme for assessing aquifer pollution vulnerability must include an element to take account of the properties of the soil zone which affect the likelihood of nutrient and pesticide leaching. Many processes causing pollutant attenuation occur at their maximum rates in this zone, as a result of its higher clay and organic content, and very much larger bacterial populations.

It must be stressed that aquifer vulnerability maps are designed to provide a general framework within which to base groundwater protection policy. They should comprise a simplified, but factual, representation of the best available scientific data on the hydrogeological environment, no more or no less. Pollution control areas may include more than one vulnerability class depending on their objective.

## *BOX 14 : Rural-Urban Competition & Conflict for Scarce Groundwater Resources in the Yemen Arab Republic*

• The water resource situation in the Yemen is extremely serious with many aquifers heavily overdrafted and population growing at close to 4% pa, leading to escalating demand in both urban centres and agricultural areas (WRAY-35, 1995).

• The situation in the area around Taiz, the third largest city, is illustrative of growing competitive pressures. Municipal water-supply (provided by NWSA) is extremely erratic, with breaks in supply often exceeding 10 days (World Bank, 1996). In response, informal water markets have evolved and many urban residents meet their needs by purchasing from tankers.

• Government is desperately seeking to improve the municipal water system and a new wellfield was constructed in the Al Hima wadi (some 25 km upstream), following negotiations with a private land owner. The area was originally swampy and generated a significant baseflow, which was utilised for agricultural irrigation.

• In the reconnaissance study groundwater resources were grossly over estimated and neither the upper alluvial aquifer nor the underlying volcanic and sandstone formations provided the sustainable yields forecast. By 1995 total groundwater production had declined to 2.6 $Mm^3/a$ from an initial yield of more than 4.0 $Mm^3/a$, and even this was at the cost of eliminating all irrigated agriculture and eradicating the lush vegetation. Most of the rural population now survive through rainfed subsistence agriculture and casual labour in Taiz. Compensation, although promised, has not yet been paid.

• In 1995, the proposal for a second emergency urban water-supply drilling program at Habeer, further upstream of Al Hima, was (not surprisingly) strongly opposed by the local rural population. Several women were shot and injured during protests aimed at stopping the drilling. Nevertheless, a new wellfield was completed in 1997 and is experiencing similar problems; compensation has been offered.

• The situation at Al Hima and Habeer contrasts sharply with the water markets through which most residents of Taiz meet their basic water needs (World Bank, 1996). Well owners adjacent to the city sell water on a daily basis, either directly to urban users or to tanker operators who retail to consumers. This informal water market is highly structured with consumers paying different rates for water of different quality (Inset). The intense and violent conflict that characterises Al Hima and Habeer is absent, and rural populations are able, at least, to increase income through water sales.

| WATER PRODUCER | WHOLESALERS/ RETAILERS | WATER CONSUMERS |
|---|---|---|
| | (indicative equivalent cost) | |
| **Well Owners** (US \$0.004/$m^3$ - pumping costs) | - | **Irrigated Agriculture** (US \$0.030/$m^3$) |
| **Well Owners** (US \$0.004/$m^3$ - pumping costs) | - | **Commercial Users (with own tankers)** (hotels, poultry farms, industrial premises, irrigated agriculture) (US \$0.140/$m^3$) |
| **Well Owners** (US \$0.004/$m^3$ - pumping costs) | **Water Tanker Operators** (US \$0.140/$m^3$) | **Smaller Users** (homeowners, hostels, restaurents, construction industry) (US \$0.210/$m^3$) |
| **Well Owners** (US \$0.004/$m^3$ - pumping costs) **Treatment Plant Operators** (with own tankers) (US \$0.004/$m^3$ - pumping costs) | **Retail Shops** (US \$1.000/$m^3$) | **Individual Users** (mainly purified drinking water) (up to US \$ 2.000/$m^3$) |

***Inset :*** *Summary of transactions and prices of informal groundwater markets in Taiz, Yemen AR*

**KEY ISSUES**

- **should government formalise and rationalise existing water markets, which cut across strongly held cultural norms on the common nature of water rights and the interpretation of some that water sale should be forbidden**
- **although such markets appear to represent a mechanism to reduce conflict and to maintain some income in rural areas, they will not address the deep-seated problem of groundwater overdraft and can impose a very high burden on the urban poor**

**Table 18: Principal hydrogeological environments and their associated pollution vulnerability**

| *Hydrogeological environment* | | *Natural travel time to saturated zone* | *Attenuation potential* | *Pollution vulnerability* |
|---|---|---|---|---|
| *Major Alluvial Formations* | unconfined | weeks-months | moderate | moderate |
| | semi-confined | years-decades | high | low |
| *Recent Coastal Limestone* | unconfined | days-weeks | low-moderate | high |
| *Inter-Montane Basins* | unconfined<br>semi-confined | years-decades | moderate | moderate |
| *Consolidated Sedimentary Aquifers* | porous sandstones | months-years | moderate | moderate |
| | karstic limestones | days-weeks | low | extreme |
| *Weathered Crystalline Basement* | unconfined | days-weeks | low-moderate | high |

*Note:* This gives a very general guide to the typical situation, and there will be much variation at local scale with detailed variations in the hydrogeology.

### *Defining Groundwater Source Protection Areas*

The objective of source protection areas (called wellhead protection zones in the USA) is to provide a special additional element of protection for selected groundwater sources (boreholes or springs). This is achieved by placing tighter controls on activities within all or part of their recharge capture area.

The outermost protection area that can be defined for an individual source is its *recharge capture area.* This is the area within which all aquifer recharge, whether derived from precipitation or surface watercourses, will be captured at the source concerned, and should not be confused with the area of hydraulic interference caused by a pumping borehole.

In practice, the definition requires further specification, and it is customary to use the maximum licensed abstraction rate together with the long-term average recharge rate when calculating such areas. It is accepted that, on this basis, the actual capture area in extreme drought will be larger than that protected. The recharge capture zones of sources are significant not only for quality protection but also in resource management terms. In situations of intensive groundwater use they could be used for aquifer exploitation control also.

In order to eliminate completely the risk of unacceptable source contamination, all potentially polluting activities would have to be prohibited or controlled to the required level within the entire recharge capture zone. This will often be untenable, due to socioeconomic pressure on land use for agriculture. Thus, some division of the recharge capture zone is required, so that generally the more severe constraints will only be applied closest to the source itself (Foster and Skinner, 1995).

This subdivision could be based on a variety of criteria, depending on the perceived pollution threat, including horizontal distance, horizontal flow time, proportion of recharge area, saturated zone dilution or attenuation capacity. The dilution and attenuation capacity of the saturated aquifer are, in practice, difficult to quantify and predict, although the latter will in a general sense increase with increasing horizontal flow distance and flow time. Intuitively, dilution might appear to be a useful criterion to delimit source protection perimeters; however, this is not necessarily so. Special protection of a proportion of the recharge area may be the preferred solution to alleviate diffuse agricultural pollution of groundwater under certain circumstances, but even then the question of which part of the recharge capture zone to protect inevitably arises.

**Table 19: Definition of aquifer vulnerability classes**

| *Integrated vulnerability class* | *Practical significance* |
|---|---|
| *Extreme* | vulnerable to most water pollutants with relatively rapid impact in many pollution scenarios |
| *High* | vulnerable to many pollutants except those highly absorbed and/or readily transformed |
| *Low* | only vulnerable to the most persistent pollutants in the very long term |
| *Negligible* | confining beds present with no significant groundwater flow |

*Note:* This overcomes those objections to the integrated vulnerability concept based on the need to specify individual contaminants.
*Source:* Foster and Skinner, 1995.

In practice an *inner protection zone* based on the distance equivalent to a specified average aquifer horizontal flow-time has been widely adopted for the prevention of pathogenic contamination of groundwater sources, from (for example) the spreading of wastewater and slurries on cultivated land. The flow-time used has varied significantly (from 10-400 days) between regulatory agencies in different countries and regions. A review of published case histories of groundwater contamination by pathogens (Lewis and others, 1982) concluded that the horizontal distance between the borehole/spring and the proven source of pollution was equivalent to no more than the distance traveled by groundwater in 20 days, despite the fact that some pathogens are capable of surviving in the subsurface for 400 days or more. A value of 50 days was thus considered a reasonably conservative basis with which to define the inner protection zone, and conforms with existing practice in many cases.

Special problems arise with the definition of recharge capture areas in situations where the groundwater divide is at a great distance, the regional hydraulic gradient is very low, and/or there are surface watercourses flowing across unconfined aquifers. A further practical complication with all source protection areas is that they vary position or have complex shapes if numerous sources are in close proximity. In the case of heavily developed aquifers, it is more practical to coalesce individual source protection zones into a larger multi-source protection area. However, if a significant proportion of the abstraction is for non-potable uses (especially irrigation) a further complication arises.

The definition of source protection areas can be achieved by using suitable computer models; provided these models are used properly they should give reliable results, within the limits of parameter uncertainty. A valuable first phase in the implementation of source protection areas is to estimate their extension and to consider their implications based on calculations using existing hydrogeological data, and it is strongly recommended that this planning exercise is undertaken by all water companies with rurally sited sources as a matter of priority.

*Undertaking Wellhead Sanitary Surveys*

While the definition of groundwater source recharge capture areas will be appropriate for higher-yielding groundwater sources used to reticulate water supply for larger villages and small towns, it is not practicable for small community and individual private domestic wells because their capture zones are very small.

In these cases, however, (as with higher-yielding potable water supply boreholes) a systematic wellhead sanitary survey is strongly recommended. A standard methodology for sanitary inspection exists (Lloyd and Helmer, 1991), in which a number of direct observations on the physical condition of the wellhead area are correlated with a fecal coliform grading derived from monitoring raw water from the source concerned (Table 20). This leads to an assessment of source contamination hazard of potentially immediate impact, and simultaneously points to appropriate risk management actions.

**Table 20: Systems of scoring for sanitary risk and confirming fecal pollution hazard for groundwater sources**

| FACTORS IN SANITARY SURVEY | | |
|---|---|---|
| **Environmental Hazards (off-site)**<br>• local caves, sinkholes or abandoned boreholes used for surface drainage or sewage disposal<br>• fissures in strata overlying water-bearing formations<br>• nearby sewers, pit latrines, cesspools, septic tanks, drains, livestock pens or farmyards<br>• nearby agricultural wastes discharged or spilled<br><br>**Construction Hazards (on-site)**<br>• well-casing leaking or not penetrated to sufficient depth, inadequate sanitary seal around casing<br>• well-casing not extended above ground or floor of pump room, or not closed at top<br>• leaks in system under vacuum<br>• wellhead, pumping plant suction pipes, or valve boxes located in pits vulnerable to flooding | | |
| scores of 4-6 indicate intermediate-to-high, and 7+ very high, potential pollution risk | | |
| **GRADE** | **FC RAW WATER COUNTS (mpn or cfu/100 ml)** | **CONFIRMED POLLUTION RISK** |
| A | 0 | none |
| B | 1-10 | low |
| C | 11-50 | intermediate-to-high |
| D | 50-1000 | high |
| E | >1000 | very high |

*Note:* The combination of simple (but clearly prescribed) visual inspection coupled with microbiological surveillance provides an effective (but low cost) approach to fecal pollution hazard assessment.
*Source:* Lloyd and Helmer, 1991.

# 5

# *THE RURAL-URBAN INTERFACE: AN ADDENDUM*

In many senses the rural-urban interface is characterized by some of the greatest groundwater resource anomalies and conflicts. It is often the area with:

- The steepest hydraulic gradient (as a result of excessive groundwater pumping in the
- Periurban environment for municipal and industrial water supply)
- The steepest water price gradients (as a result of the variation in groundwater abstraction charges and end-user values between the urban and rural environment)
- The heaviest subsurface contaminant load and greatest risk of groundwater pollution (as a result of periurban industrial development and intensification of agricultural production by horticulture to meet urban demands).

It is not the intention in this addendum to enter into detailed discussion of urban groundwater issues, nor the evolution of groundwater exploitation in urban and periurban areas, since this was dealt with in an earlier companion World Bank Technical Paper on "Groundwater in Urban Development" (Foster and others, 1998). Discussion here, therefore, is restricted to consideration of those issues which most impact upon the status of groundwater resources and/or the rural community themselves:

- Competition for available groundwater resources between urban and rural users, resulting from the pressure to transfer water supplies to neighboring urban areas
- Potential constraints imposed on the agricultural community by policies aimed at protecting potable groundwater quality in the vicinity of municipal wellfields
- The potential impact on potable groundwater quality of the reuse of urban wastewater for agricultural irrigation.

## Groundwater Resource Competition and Transfers

Many regimes of land and water resource administration permit municipal water utilities/companies to explore for and develop new groundwater supplies well beyond current urban limits in contiguous agricultural areas. In some instances the impact of major wellfield development for the rural community can include:

- Increased pumping head and energy costs, or even the need to reset/redimension/replace pumping plant and to deepen boreholes, for the owners and operators of irrigation wells
- Increased rates of aquifer overdraft in situations of resource scarcity, compromising further the long-term sustainability of groundwater resources.

In other situations the increased pressure of groundwater resources will come from private abstractors providing urban services, including the provision of tankered water supplies.

The situation is frequently further complicated by inadequate characterization of the local groundwater system, and misconceptions, for example, about the degree of hydraulic independence of deeper aquifers under exploitation for urban and industrial supplies, and shallower aquifers providing the water supply for agricultural irrigation.

All too often there is inadequate investigation of the potential impacts of new urban wellfield developments on existing agricultural groundwater users. Moreover, there is no acceptance of the concept of paying compensation for interference with pre-existing water rights nor existence of a transparent system by which such compensation should be estimated.

On the other side, there is often a long history of not levying any charge or realistic charge for groundwater exploitation for irrigation, leading to an entrenched situation as regards the undervaluation of groundwater resources and consequently their inefficient use in agriculture.

An example of the level of resource conflict that can arise, and potential solutions in terms of resource management, is given in Box 14. What is apparent is that the way out of the more deeply entrenched rural-urban groundwater resource conflicts lies in establishing and registering groundwater abstraction rights and then using economic mechanisms to constrain and to allocate available resources more effectively. The latter may involve greatly increased abstraction charges to reflect resource scarcity, but it may be easier to introduce economic measures by establishing water rights markets for which the regulatory agency acts as broker.

## Municipal Wellfield Protection Issues

Another potential dimension of the urban-rural groundwater resource conflict is the pressure that may arise for land-use controls in the vicinity of urban wellfields. A rational component of the development of a new municipal wellfield is to mobilize actions to protect the asset: first through the definition of a protection zone corresponding to all or part of the source capture area and second by controls on land-use activity within the protection zone according to the pollution vulnerability of the aquifer system involved (Box 13).

Sooner or later it may be recognized that some control over the application of agricultural fertilizers, pesticides, and slurries, or on livestock grazing densities may, in fact, be needed to protect the potability of the groundwater supply. In extreme cases more radical changes in cultivation regime may be sought. Where the latter involves actual land purchase by the municipal water company, an element of compensation to those individuals in the agricultural sector is implicit. However, where constraints in agriculture are imposed following representations to the regulatory agency, conflicts may arise.

This type of land-use/water quality interaction issue is, as yet, far from finding adequate institutional resolution, and introduces potential inequities between neighboring farmers in a relatively small land area. The question of compensation being paid to affected farmers arises, and whether the revenue should be raised from charges imposed on the municipal water-users. There is no need for the regulatory agency to act as a broker in this respect, but it is useful for the brokerage system to involve the regulator as registrar.

## Urban Wastewater Reuse for Irrigation

Where cities have significant cover of main sewerage (as opposed to in-situ sanitation), substantial volumes of wastewater are continuously discharged, normally close to the downstream urban-rural interface. This wastewater represents both an important water resource for irrigation (the only one worldwide which is growing in volume and availability) and also a potential public health hazard, unless the WHO 1989 Guidelines for Wastewater Reuse are respected. The level of treatment varies but rarely extends beyond primary settlement. Even where it is more complete the objective is normally to reduce environmental impact in receiving watercourses (where BOD, SS and P are the main considerations) rather than the elimination of pathogens and nitrate load. As a result wastewater still normally has a major potential to pollute underlying aquifers if the local streams/rivers are influent (infiltrate to groundwater).

In climates which have an extended dry season or are generally arid, urban wastewater often provides the bulk of riverflow downstream of major conurbations for many months in the year and is likely to be used for irrigation of agricultural crops on downstream alluvial tracts (Box 15). Indeed, some urban water utilities are in the process of offering partially-treated wastewater and financing improvements in irrigation technology to farmers in exchange for groundwater abstraction rights to reduce periurban groundwater resource competition and overdraft.

However, given the normally high suspended solids and organic matter content of wastewater, application is by flood irrigation and results in high rates of infiltration to groundwater on permeable alluvial-terrace soils (Foster and others, 1994). The degree of groundwater pollution hazard involved varies widely with the aquifer pollution vulnerability and the characteristics of the wastewater (especially its salinity and content of toxic organic chemicals and heavy metals).

At the same time it must be noted that it is possible to use some wastewaters for groundwater recharge, effecting tertiary-level treatment by infiltration through the vadose zone. This process is capable of producing groundwater of sufficient quality to allow safe irrigation of high-value horticultural crops subsequently. However, the infiltration process will not alone regenerate water of potable quality since various contaminants such as nitrates and synthetic organic (community and industrial) chemicals at least will persist and only be reduced by dilution. In less favorable circumstances there may also be residual contamination by some fecal pathogens, excessive salinity and/or other chemicals.

Thus, while wastewater reuse is much needed in the urban-rural interface and around major conurbations. At whatever level it is practiced there is a need for:

- Careful planning
- Operational control
- Systematic monitoring.

At present, it rarely receives adequate attention. All too often it is practiced on an anarchical basis which threatens the well-being of agricultural workers, the health of those consuming their products and the long-term quality of groundwater in the underlying aquifer which may be an important source of potable water supply. The larger is the wastewater irrigation area, the proportion of wastewater to freshwater in the area and the salinity of the wastewater itself, the greater will be the overall impact on groundwater quality and the potential problem of locating and protecting groundwater supplies of potable quality in the area concerned.

## *BOX 15 : Wastewater Re-Use for Agricultural Irrigation in Central Mexico: Benefits, Problems and Solutions*

• The city of León-Guanajuato (population 1.2 million) is one of the fastest growing cities in México, and is highly dependent on groundwater for public supply. Groundwater is abstracted mainly from aquifers downstream, including areas where city wastewaters are used for agricultural irrigation. León wastewater is of relatively high salinity and chromium content because of the major leather processing and shoe manufacturing industry.

• A recent study showed (Foster, 1996; Chilton et al, 1998) that high rates of recharge from excess wastewater irrigation on alfalfa and maize southwest of the city (coupled with no agricultural abstraction) have helped maintain groundwater levels within 10m depth, despite intensive abstraction from deeper horizons for municipal water supply. In adjacent areas water levels are falling at 2-5m/a.

• However, salinity problems are beginning to affect a number of production wells in the wastewater irrigated area. In the most seriously affected well, the chloride concentration rose from 100 mg/l to 230 mg/l in 2 years (even though the boreholes in this wellfield are screened from 200-400m depth) and it is predicted that they could rise to 400mg/l by 2010 in all the neighbouring wells if no remedial action is taken. There is also evidence of increasing nitrate concentrations.

• In contrast, although the wastewater also contains large concentrations of chromium salts, Cr concentrations in groundwater remain low. Soil sampling has confirmed that chromium and other heavy metals are accumulating in the soil, with very little passing below a depth of 0.3m. Neither are significant levels of pathogenic micro-organisms or fecal coliform indicators found in the groundwater.

• It is thus not necessarily the most toxic component of an effluent which poses the main threat to groundwater, and this example highlights the importance of understanding pollutant transport in the subsurface. Future management therefore needs to address the problem of rising salinity, while trying to continue to maximise the reuse of wastewater in agriculture.

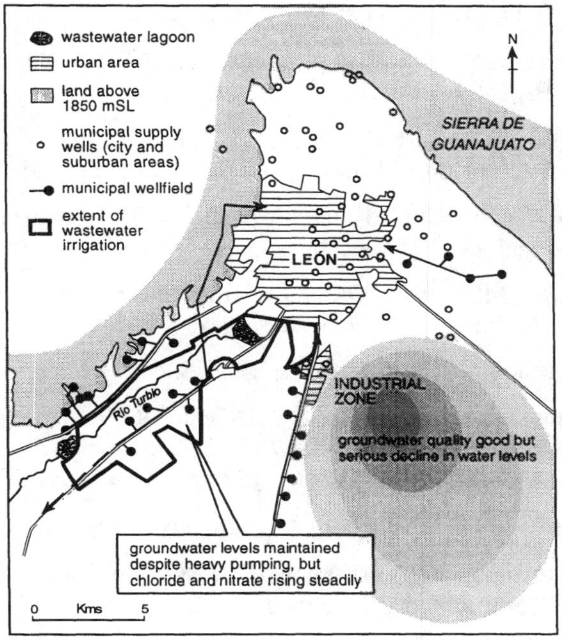


**Inset:** *Location of municipal wellfields and wastewater reuse area of León-Guanajuato, Mexico.*

### KEY ISSUES:

- **address rising salinity (more urgently than conventional wastewater treatment) through separate collection/treatment of saline industrial effluents (although substantial time-lag before benefits felt as improved groundwater quality)**
- **shallow groundwater pumping for irrigation in existing wastewater reuse area to intercept and recycle saline recharge; this may have implications for agricultural production and soil fertility, and will also imply extending reuse area**
- **remove affected municipal production wells from supply to reduce downward leakage of saline recharge, which will also require demand management measures (mains leakage control and private use constraints) in view of reduced supply**

# *REFERENCES*

**(Key works in relation to the development and perusal of this paper are in bold type)**


Ahmad, K. G. 1990. "Rehabilitation of Deteriorated Tubewells in Punjab SCARPs." *Water Wells: Monitoring, Maintenance and Rehabilitation*. London: Spon. 291-302.

Arlosoroff, S, Tschanneri G., Grey D. R. C., Journey W., Karp A., Langeneggar O., and Roche R. 1987. "Community Water Supply: The Handpump Option." UNDP-World Bank Water and Sanitation Program Publn. World Bank, Washington, D.C.

Ayers, R. S., and Westcot D. W. 1985. "Water Quality for Agriculture." UN-FAO Irrigation and Drainage Paper 29.

Benson, C., And Clay E. 1998. "The Impact of Drought on Sub-Saharan Economies: A Preliminary Examination." World Bank Technical Paper 401. World Bank, Washington, D.C.

BGS-IH-MLWDZ. 1996. "Community Gardens Using Limited Groundwater Resources." Pamphlet. British Geological Survey, Nottingham.

BGS-MMD. 1998. "Groundwater Studies of Arsenic Contamination in Bangladesh." *BGS Special Report* (5 Vols) British Geological Survey, Nottingham.

Bredehoeft, J. D. 1997. "Safe Yield And The Water Budget Myth." *Ground Water* 35: 929-934.

Calow, R. C., Robins N. S., Macdonald A. M., Macdonald D. M. J., Gibbs B. R., Orpen W. R. G., Mtembezeka P., Andrews A. J., and Appioh S. O. 1997. "Groundwater Management in Drought Prone Areas of Africa." *Intl J Water Res Dev* 13: 241-261.

Cgwb. 1995. Special Issue On Groundwater Statistics. *Bhu-Jal News*: 1-162.

**Chambers, R., and Shah T. 1989. *To The Land Of The Poor—Water And Trees*. New Delhi: Commonwealth Publishers.**

Chilton, P. J. 1995. "Salinization of Soils and Aquifers." BGS Technical Report WD/95/26: 54-64. British Geological Survey, Nottingham.

**Chilton, P. J., and Foster S. S. D. 1995. "Hydrogeological Characterization and Water Supply Potential of Basement Aquifers in Tropical Africa." *Hydrogeol Journal* 3: 36-49.**

Chilton, P. J., Vlugman A. A., and Foster S. S. D. 1990. "A Groundwater Pollution Risk Assessment for Public Water Supply Sources in Barbados." AWRA Intl Conf Tropical Hydrology and Caribbean Water Resources (San Juan De Puerto Rico, 1990): 279-288.

Chilton, P. J., Stuart M. E., Escolero O., Marks R. J., Gonzalez A., and Milne C. J. 1998. "Groundwater Recharge and Pollutant Transport beneath Wastewater Irrigation: The Case of Leon, Mexico." Geol Soc Spec Publn 130: 153-168.

Clarke, R., Lawrence A. R., and Foster S. S. D. 1996. "Groundwater—A Threatened Resource." UNEP Environment Library Series 15.

Collin, J. J., and Margat J. 1993. "Overexploitation of Water Resources: Overreaction or Economic Reality?" *Hydroplus* 36: 26-37.

**Custodio, E. 1992. "Hydrogeological and Hydrochemical Aspects of Aquifer Overexploitation."** ***IAH Hydrogeology Selected Papers*** **35: 3-28.**

Dfid-Well. 1998. *Guidance Manual on Water Supply and Sanitation Programs.* London: Department For International Development.

Dhawan, B. D. 1995. *Groundwater Depletion, Land Degradation And Irrigated Agriculture In India.* New Delhi: Commonwealth Publishers.

Dinar, A., and Subramanian A. 1997. "Water Pricing Experiences: An International Perspective." World Bank Technical Paper 396. World Bank, Washington, D.C.

Driscoll, F. G. 1986. *Groundwater And Wells.* St. Paul, Minnesota: Johnson Division.

Edmunds, W. M., and Smedley P. L. 1995. "Groundwater Geochemistry and Health: An Overview." Geol *Soc Spec Publn* 113: 91-105. (Also as British Geological Survey/Nottingham Pamphlet.)

Farr, J. L., Spray P. R., and Foster S. S. D. 1982. "Groundwater Supply Exploration in Semi-Arid Regions for Livestock Extension—A Technical and Economic Appraisal." *Water Supply And Management* 6: 343-353.

Feitelson, E., and Haddad M. 1998. "Identification of Joint Management Structures for Shared Aquifers." World Bank Technical Paper 415. World Bank, Washington, D.C.

Foster, S. S. D. 1989. "Diffuse Pollution of Groundwater by Agriculture: Lessons Learnt and Future Prospects." Aih Intl Conf Recent Advances In Groundwater Hydrogeology (Tampa/1988): 185-194.

**Foster, S. S. D. 1992. "Unsustainable Development and Irrational Exploitation of Groundwater Resources in Developing Nations—An Overview."** ***IAH Hydrogeology Selected Papers*** **3: 321-336.**

———. 1993. "Groundwater Conditions and Problems Characteristic of The Humid Tropics." IAHs Publn 216: 433-449.

———. 1996. "Groundwater Quality Concerns in Rapidly Developing Cities." Aih Intl Conf Hydrology and Hydrogeology Of Urban Areas (Boston/1996) Miu 12-26.

**Foster, S. S. D., and Chilton P. J. 1998. "As The Land So The Water—The Effects of Agricultural Cultivation on Groundwater." UNESCO-CIHEAM Seminar Publn** ***Agricultural Threats To Groundwater Quality*** **(Zaragoza/1996): 15-43.**

Foster, S. S. D., Gale I. N., and Hespanhol I. 1994. "Impacts of Wastewater Use and Disposal on Groundwater." BGS Technical Report WD/94/55. British Geological Survey, Nottingham.

Foster, S. S. D., and Hirata R. C. A. 1988. "Groundwater Pollution Risk Assessment: A Methodology Using Available Data." WHO-PAHO-CEPIS Publn. CEPIS/Lima.

**Foster, S. S. D., and Lawrence A. R. 1995. "Groundwater Quality in Asia: An Overview of Trends and Concerns."** ***UN-ESCAP Water Resources Journal*** **Series C, 184: 97-110.**

**Foster, S. S. D., Lawrence A. R., and Morris B. L. 1998. "Groundwater in Urban Development: Assessing Management Needs and Formulating Policy Strategies." World Bank Technical Paper 390. World Bank, Washington, D.C.**

**Foster, S. S. D., and Skinner A. C. 1995. "Groundwater Protection: The Science and Practice of Land Surface Zoning." IAHS Publn 225: 471-482.**

Freeze, R. A., and Cherry J. A. 1979. *Groundwater.* New Jersey: Prentice-Hall.

Howsam, P. 1995. "Groundwater Abstraction—Monitoring, Operation and Maintenance." BGS Technical Report WD/95/26: 1-9. British Geological Survey, Nottingham.

**Kahnert, F., and Levine G. 1993. "Groundwater Irrigation and The Rural Poor—Options for Development in The Gangetic Basin." World Bank Colloquium Publn *Groundwater Irrigation, 1989.* Washington, D.C.:The World Bank.**

Koehler, C. 1995. "Water Rights and The Public Trust Doctrine: Resolution of The Mono Lake." *Ecol Law Quart* 22: 511-589.

Kolpin, D. W., and Goolsby D. A. 1995. "A Regional Monitoring Network to Investigate the Occurrence of Agricultural Chemicals in Near-Surface Aquifers of the Midcontinental USA." IAHS Pubn 225 : 13-20.

Korom, S. F. 1992. "Natural Denitrification in the Saturated Zone: A Review." *Water Res Research* 28: 1657-1668.

Kromm, D. E., and White S. E. 1990. "Conserving Water in the High Plains." Kansas State University Spec Publn.

Langenegger, O. 1994. "Groundwater Quality and Handpump Corrosion in West Africa." UNDP-World Bank Water And Sanitation Program Report 8. World Bank, Washington, D.C.

Lawrence, A. R., and Foster S. S. D. 1986. "Denitrification in a Limestone Aquifer in Relation to the Security of Low Nitrate Groundwater Supplies." *J Inst Water Engrs And Scientists* 40: 159-172.

Llamas, M. R. 1992. "La Surexploitation des Aquiferes: Aspects Techniques et Institutionneis." *Hydrogeologie* 4: 139-144.

Llamas, M. R. 1998. "Groundwater Overexploitation." Unesco Congress Water In 21st Century: A Looming Crisis (Paris/1998) 2:1-20.

Lerner, D. N., Issar A. S., and Simmers I. 1990. "Groundwater Recharge." IAH Intl Contrbns Hydrogeology 8.

Lewis, W. J., and Chilton P. J. 1984. "Performance of Sanitary Completion Measures of Wells and Boreholes Used for Rural Water Supplies in Malawi." IAHS Publn 144: 235-247.

Lewis, W. J., Foster S. S. D., and Drasar B. S. 1982. "The Risk of Groundwater Pollution by On-Site Sanitation in Developing Countries." WHO-IRCWD Report 01/82. IRCWD/Dubendorf.

Lloyd, J. W. 1997. "The Future Use of Aquifers in Water Resources Management in Arid Areas." *Arabian J Sci Eng* 22: 33-45.

Lloyd, B., and Helmer R. 1991. "Surveillance of Drinking Water Quality in Rural Areas." WHO-UNEP Publn. London: Longman.

**Moench, M. 1994. "Approaches to Groundwater Management: To Control or Enable." *Economic and Political Weekly* (24 September 1994) A135-A146.**

**NRC. 1997 *Valuing Groundwater: Economic Concepts and Approaches.* Washington, D.C.: National Academy Press.**

Prove, B. G., Mcshane T. J., Reghenzani J. R., Armour J. D., Sen S., and Moody P. W. 1994. "Nutrient Loss via Drainage from the Major Agricultural Industries on the Wet Tropical Coast of Queensland." Proc IAH Congress Water Down Under (Adelaide/1994): 439-443.

Postel, S. 1999. *Pillar of Sand: Can the Irrigation Miracle Last?* Worldwatch Institute Book. New York: Norton.

Reisner, M., and Bates S. 1990. *Overtapped Oasis.* Washington, D.C.: Island Press

**Salman, S. M. A. 1999. "Groundwater: Legal and Policy Perspectives." World Bank Technical Paper 456.** World Bank, Washington D.C.

Scharp, C. 1994. "Groundwater Protection for the Managua Aquifer—Development of a Planning Tool." IAHS Publn 222: 443-451.

Scharp, C., Alveteg T., and Johansson P. O. 1997. "Assigning a Groundwater Protection Value: Methodology Development." Proc IAH Congress Groundwater In The Urban Environment (Nottingham/1997) I: 659-664.

**Schiffler, M. 1998. *The Economics of Groundwater Management in Arid Countries.* GDI Book Series II. London: Frank Cass.**

**Shah, T. 1993. *Groundwater Markets and Irrigation Development: Political Economy and Practical Policy.* Oxford: Oxford University Press.**

Simmers, I., Hendriks J. M. H., Kruseman G. P., and Rushton K. R. 1997. "Recharge of Phreatic Aquifers in Semi-Arid Regions." IAH Intl Contrbns Hydrogeology 19.

Sophocleous, M. 1997. "Managing Water Resources Systems: Why "Safe-Yield" is Not Sustainable." *Ground Water* 35: 361-365.

Spalding, R. F., and Exner M. 1993. "Occurrence of Nitrate in Groundwater—A Review." *J Environ Quality* 22: 392-402.

Subramanian, A., Jagannathan N. V., and Meinzen-Dick R. 1997. "User Organizations for Sustainable Water Services." World Bank Technical Paper 354. World Bank, Washington, D.C.

Turton, A. R. 1999. "Water Scarcity and Social Adaptive Capacity: Towards an Understanding of the Social Dynamics of Managing Water Scarcity in Developing Countries." SOAS Water Issues Study Group Occ Paper 9. (University of London).

UNDP/PROWESS. 1990. *Taking the Pulse of Community Management in Water and Sanitation.* New York: UNDP.

UN-FAO. 1997. "Seawater Intrusion in Coastal Aquifers: Guidelines for Study, Monitoring and Control." UN-FAO Water Reports II.

**Van Dongen, P., and Woodhouse M. 1994. "Finding Groundwater: A Project Managers Guide to Techniques and How to Use Them." UNDP-World Bank Water and Sanitation Program Publn. World Bank, Washington, D.C.**

Vrba, J., and Romijn E. 1986. "Impact of Agricultural Activities on Groundwater." IAH Intl Contribns Hydrogeology 5.

WHO-IWSC. 1993. "Management of Operation and Maintenance in Rural Drinking Water Supply and Sanitation." WHO/CWS Publn: 92.16. WHO, Geneva.

Wood, B. P., and Chilton P. J. 1995. "Occurrence and Distribution of Ametryne, Atrazine and its Deetheylated Metabolite in Barbados Groundwater." BGS Technical Report WD/95/26: 123-129. British Geological Survey, Nottingham.

World Bank. 1994. *Water Resources Policy.* Washington, D.C.: World Bank.

———. 1996. "Yemen-Tairz Water Supply Project." World Bank Report 15599. World Bank, Washington, D.C.

———. 1997. "Jordan Water Sector Review." World Bank Report 17095. World Bank, Washington, D.C.

———. 1998-a. "India Water Resources Management Sector Review: Groundwater Regulation and Management Report." World Bank, Washington, D.C.

———. 1998-b. "Mexico Water Resources Management Project." World Bank Report 15435. World Bank Report, Washington, D.C.

WRAY-35 1995. "The Water Resources of Yemen." Republic of Yemen—General Department of Hydrology/TNO Institute of Applied Geoscience Report. Delft, The Netherlands.

Wright, E. P., and Burgess W. G. 1992. "The Hydrogeology of the Crystalline Basement Aquifer in Africa." Geol Soc Spec Publn 66.

Young, R. A. 1992. "Managing Aquifer Over-Exploitation: Economics and Policies." *IAH Hydrogeology Selected Papers* 3: 199-222.

Young, R. A. 1996. "Measuring Economic Benefits for Water Investments and Policies." World Bank Technical Paper 338. World Bank, Washington, D.C.

# Distributors of World Bank Group Publications

Prices and credit terms vary from country to country. Consult your local distributor before placing an order.

**ARGENTINA**
World Publications SA
Av. Cordoba 1877
1120 Ciudad de Buenos Aires
Tel: (54 11) 4815-8156
Fax: (54 11) 4815-8156
E-mail: wpbooks@infovia.com.ar

**AUSTRALIA, FIJI, PAPUA NEW GUINEA, SOLOMON ISLANDS, VANUATU, AND SAMOA**
D.A. Information Services
648 Whitehorse Road
Mitcham 3132, Victoria
Tel: (61) 3 9210 7777
Fax: (61) 3 9210 7788
E-mail: service@dadirect.com.au
URL: http://www.dadirect.com.au

**AUSTRIA**
Gerold and Co.
Weihburggasse 26
A-1011 Wien
Tel: (43 1) 512-47-31-0
Fax: (43 1) 512-47-31-29
URL: http://www.gerold.co/at.online

**BANGLADESH**
Micro Industries Development Assistance Society (MIDAS)
House 5, Road 16
Dhanmondi R/Area
Dhaka 1209
Tel: (880 2) 326427
Fax: (880 2) 811188

**BELGIUM**
Jean De Lannoy
Av. du Roi 202
1060 Brussels
Tel: (32 2) 538-5169
Fax: (32 2) 538-0841

**BRAZIL**
Publicacões Tecnicas Internacionais Ltda.
Rua Peixoto Gomide, 209
01409 Sao Paulo, SP.
Tel: (55 11) 259-6644
Fax: (55 11) 258-6990
E-mail: postmaster@pti.uol.br
URL: http://www.uol.br

**CANADA**
Renouf Publishing Co. Ltd.
5369 Canotek Road
Ottawa, Ontario K1J 9J3
Tel: (613) 745-2665
Fax: (613) 745-7660
E-mail: order.dept@renoufbooks.com
URL: http:// www.renoufbooks.com

**CHINA**
China Financial & Economic Publishing House
8, Da Fo Si Dong Jie
Beijing
Tel: (86 10) 6401-7365
Fax: (86 10) 6401-7365

China Book Import Centre
P.O. Box 2825
Beijing

Chinese Corporation for Promotion of Humanities
52, You Fang Hu Tong,
Xuan Nei Da Jie
Beijing
Tel: (86 10) 660 72 494
Fax: (86 10) 660 72 494

**COLOMBIA**
Infoenlace Ltda.
Carrera 6 No. 51-21
Apartado Aereo 34270
Santafé de Bogotá, D.C.
Tel: (57 1) 285-2798
Fax: (57 1) 285-2798

**COTE D'IVOIRE**
Center d'Edition et de Diffusion Africaines (CEDA)
04 B.P. 541
Abidjan 04
Tel: (225) 24 6510; 24 6511
Fax: (225) 25 0567

**CYPRUS**
Center for Applied Research
Cyprus College
6, Diogenes Street, Engomi
P.O. Box 2006
Nicosia
Tel: (357 2) 59-0730
Fax: (357 2) 66-2051

**CZECH REPUBLIC**
USIS, NIS Prodejna
Havelkova 22
130 00 Prague 3
Tel: (420 2) 2423 1486
Fax: (420 2) 2423 1114
URL: http://www.nis.cz/

**DENMARK**
SamfundsLitteratur
Rosenoerns Allé 11
DK-1970 Frederiksberg C
Tel: (45 35) 351942
Fax: (45 35) 357822
URL: http://www.sl.cbs.dk

**ECUADOR**
Libri Mundi
Libreria Internacional
P.O. Box 17-01-3029
Juan Leon Mera 851
Quito
Tel: (593 2) 521-606; (593 2) 544-185
Fax: (593 2) 504-209
E-mail: librimu1@librimundi.com.ec
E-mail: librimu2@librimundi.com.ec

**CODEU**
Ruiz de Castilla 763, Edif. Expocolor
Primer piso, Of. #2
Quito
Tel/Fax: (593 2) 507-383; 253-091
E-mail: codeu@impsat.net.ec

**EGYPT, ARAB REPUBLIC OF**
Al Ahram Distribution Agency
Al Galaa Street
Cairo
Tel: (20 2) 578-6083
Fax: (20 2) 578-6833

The Middle East Observer
41, Sherif Street
Cairo
Tel: (20 2) 393-9732
Fax: (20 2) 393-9732

**FINLAND**
Akateeminen Kirjakauppa
P.O. Box 128
FIN-00101 Helsinki
Tel: (358 0) 121 4418
Fax: (358 0) 121-4435
E-mail: akatilaus@stockmann.fi
URL: http://www.akateeminen.com

**FRANCE**
Editions Eska; DBJ
48, rue Gay Lussac
75005 Paris
Tel: (33-1) 55-42-73-08
Fax: (33-1) 43-29-91-67

**GERMANY**
UNO-Verlag
Poppelsdorfer Allee 55
53115 Bonn
Tel: (49 228) 949020
Fax: (49 228) 217492
URL: http://www.uno-verlag.de
E-mail: unoverlag@aol.com

**GHANA**
Epp Books Services
P.O. Box 44
TUC
Accra
Tel: 223 21 778843
Fax: 223 21 779099

**GREECE**
Papasotiriou S.A.
35, Stournara Str.
106 82 Athens
Tel: (30 1) 364-1826
Fax: (30 1) 364-8254

**HAITI**
Culture Diffusion
5, Rue Capois
C.P. 257
Port-au-Prince
Tel: (509) 23 9260
Fax: (509) 23 4858

**HONG KONG, CHINA; MACAO**
Asia 2000 Ltd.
Sales & Circulation Department
302 Seabird House
22-28 Wyndham Street, Central
Hong Kong, China
Tel: (852) 2530-1409
Fax: (852) 2526-1107
E-mail: sales@asia2000.com.hk
URL: http://www.asia2000.com.hk

**HUNGARY**
Euro Info Service
Margitszgeti Europa Haz
H-1138 Budapest
Tel: (36 1) 350 80 24, 350 80 25
Fax: (36 1) 350 90 32
E-mail: euroinfo@mail.matav.hu

**INDIA**
Allied Publishers Ltd.
751 Mount Road
Madras - 600 002
Tel: (91 44) 852-3938
Fax: (91 44) 852-0649

**INDONESIA**
Pt. Indira Limited
Jalan Borobudur 20
P.O. Box 181
Jakarta 10320
Tel: (62 21) 390-4290
Fax: (62 21) 390-4289

**IRAN**
Ketab Sara Co. Publishers
Khaled Eslamboli Ave., 6th Street
Delafrooz Alley No. 8
P.O. Box 15745-733
Tehran 15117
Tel: (98 21) 8717819; 8716104
Fax: (98 21) 8712479
E-mail: ketab-sara@neda.net.ir

Kowkab Publishers
P.O. Box 19575-511
Tehran
Tel: (98 21) 258-3723
Fax: (98 21) 258-3723

**IRELAND**
Government Supplies Agency
Oifig an tSoláthair
4-5 Harcourt Road
Dublin 2
Tel: (353 1) 661-3111
Fax: (353 1) 475-2670

**ISRAEL**
Yozmot Literature Ltd.
P.O. Box 56055
3 Yohanan Hasandlar Street
Tel Aviv 61560
Tel: (972 3) 5285-397
Fax: (972 3) 5285-397

R.O.Y. International
PO Box 13056
Tel Aviv 61130
Tel: (972 3) 649 9469
Fax: (972 3) 648 6039
E-mail: royil@netvision.net.il
URL: http://www.royint.co.il

Palestinian Authority/Middle East Index Information Services
P.O.B. 19502 Jerusalem
Tel: (972 2) 6271219
Fax: (972 2) 6271634

**ITALY, LIBERIA**
Licosa Commissionaria Sansoni SPA
Via Duca Di Calabria, 1/1
Casella Postale 552
50125 Firenze
Tel: (39 55) 645-415
Fax: (39 55) 641-257
E-mail: licosa@ftbcc.it
URL: http://www.ftbcc.it/licosa

**JAMAICA**
Ian Randle Publishers Ltd.
206 Old Hope Road, Kingston 6
Tel: 876-927-2085
Fax: 876-977-0243
E-mail: irpl@colis.com

**JAPAN**
Eastern Book Service
3-13 Hongo 3-chome, Bunkyo-ku
Tokyo 113
Tel: (81 3) 3818-0861
Fax: (81 3) 3818-0864
E-mail: orders@svt-ebs.co.jp
URL: http://www.bekkoame.or.jp/~svt-ebs

**KENYA**
Africa Book Service (E.A.) Ltd.
Quaran House, Mfangano Street
P.O. Box 45245
Nairobi
Tel: (254 2) 223 641
Fax: (254 2) 330 272

Legacy Books
Loita House
Mezzanine 1
P.O. Box 68077
Nairobi
Tel: (254) 2-330853, 221426
Fax: (254) 2-330854, 561654
E-mail: Legacy@form-net.com

**KOREA, REPUBLIC OF**
Dayang Books Trading Co.
International Division
783-20, Pangba Bon-Dong,
Socho-ku
Seoul
Tel: (82 2) 536-9555
Fax: (82 2) 536-0025
E-mail: seamap@chollian.net

Eulyoo Publishing Co., Ltd.
46-1, Susong-Dong
Jongro-Gu
Seoul
Tel: (82 2) 734-3515
Fax: (82 2) 732-9154

**LEBANON**
Librairie du Liban
P.O. Box 11-9232
Beirut
Tel: (961 9) 217 944
Fax: (961 9) 217 434
E-mail: hsayegh@librairie-du-liban.com.lb
URL: http://www.librairie-du-liban.com.lb

**MALAYSIA**
University of Malaya Cooperative Bookshop, Limited
P.O. Box 1127
Jalan Pantai Baru
59700 Kuala Lumpur
Tel: (60 3) 756-5000
Fax: (60 3) 755-4424
E-mail: umkoop@tm.net.my

**MEXICO**
INFOTEC
Av. San Fernando No. 37
Col. Toriello Guerra
14050 Mexico, D.F.
Tel: (52 5) 624-2800
Fax: (52 5) 624-2822
E-mail: infotec@rtn.net.mx
URL: http://rtn.net.mx

Mundi-Prensa Mexico S.A. de C.V.
c/Rio Panuco, 141-Colonia Cuauhtemoc
06500 Mexico, D.F.
Tel: (52 5) 533-5658
Fax: (52 5) 514-6799

**NEPAL**
Everest Media International Services (P.) Ltd.
GPO Box 5443
Kathmandu
Tel: (977 1) 416 026
Fax: (977 1) 224 431

**NETHERLANDS**
De Lindeboom/Internationale Publicaties b.v.-
P.O. Box 202, 7480 AE Haaksbergen
Tel: (31 53) 574-0004
Fax: (31 53) 572-9296
E-mail: lindeboo@worldonline.nl
URL: http://www.worldonline.nl/~lindeboo

**NEW ZEALAND**
EBSCO NZ Ltd.
Private Mail Bag 99914
New Market
Auckland
Tel: (64 9) 524-8119
Fax: (64 9) 524-8067

Oasis Official
P.O. Box 3627
Wellington
Tel: (64 4) 499 1551
Fax: (64 4) 499 1972
E-mail: oasis@actrix.gen.nz
URL: http://www.oasisbooks.co.nz/

**NIGERIA**
University Press Limited
Three Crowns Building Jericho
Private Mail Bag 5095
Ibadan
Tel: (234 22) 41-1356
Fax: (234 22) 41-2056

**PAKISTAN**
Mirza Book Agency
65, Shahrah-e-Quaid-e-Azam
Lahore 54000
Tel: (92 42) 735 3601
Fax: (92 42) 576 3714

Oxford University Press
5 Bangalore Town
Sharae Faisal
PO Box 13033
Karachi-75350
Tel: (92 21) 446307
Fax: (92 21) 4547640
E-mail: ouppak@TheOffice.net

Pak Book Corporation
Aziz Chambers 21, Queen's Road
Lahore
Tel: (92 42) 636 3222; 636 0885
Fax: (92 42) 636 2328
E-mail: pbc@brain.net.pk

**PERU**
Editorial Desarrollo SA
Apartado 3824, Ica 242 OF. 106
Lima 1
Tel: (51 14) 285380
Fax: (51 14) 286628

**PHILIPPINES**
International Booksource Center Inc.
1127-A Antipolo St, Barangay, Venezuela
Makati City
Tel: (63 2) 896 6501; 6505; 6507
Fax: (63 2) 896 1741

**POLAND**
International Publishing Service
Ul. Piekna 31/37
00-677 Warzawa
Tel: (48 2) 628-6089
Fax: (48 2) 621-7255
E-mail: books%ips@ikp.atm.com.pl
URL: http://www.ipscg.waw.pl/ips/export

**PORTUGAL**
Livraria Portugal
Apartado 2681, Rua Do Carm o 70-74
1200 Lisbon
Tel: (1) 347-4982
Fax: (1) 347-0264

**ROMANIA**
Compani De Librarii Bucuresti S.A.
Str. Lipscani no. 26, sector 3
Bucharest
Tel: (40 1) 313 9645
Fax: (40 1) 312 4000

**RUSSIAN FEDERATION**
Isdatelstvo <Ves Mir>
9a, Kolpachniy Pereulok
Moscow 101831
Tel: (7 095) 917 87 49
Fax: (7 095) 917 92 59
ozimarin@glasnet.ru

**SINGAPORE; TAIWAN, CHINA MYANMAR; BRUNEI**
Hemisphere Publication Services
41 Kallang Pudding Road #04-03
Golden Wheel Building
Singapore 349316
Tel: (65) 741-5166
Fax: (65) 742-9356
E-mail: ashgate@asianconnect.com

**SLOVENIA**
Gospodarski vestnik Publishing Group
Dunajska cesta 5
1000 Ljubljana
Tel: (386 61) 133 83 47; 132 12 30
Fax: (386 61) 133 80 30
E-mail: repansekj@gvestnik.si

**SOUTH AFRICA, BOTSWANA**
For single titles:
Oxford University Press Southern Africa
Vasco Boulevard, Goodwood
P.O. Box 12119, N1 City 7463
Cape Town
Tel: (27 21) 595 4400
Fax: (27 21) 595 4430
E-mail: oxford@oup.co.za

For subscription orders:
International Subscription Service
P.O. Box 41095
Craighall
Johannesburg 2024
Tel: (27 11) 880-1448
Fax: (27 11) 880-6248
E-mail: iss@is.co.za

**SPAIN**
Mundi-Prensa Libros, S.A.
Castello 37
28001 Madrid
Tel: (34 91) 4 363700
Fax: (34 91) 5 753998
E-mail: libreria@mundiprensa.es
URL: http://www.mundiprensa.com/

Mundi-Prensa Barcelona
Consell de Cent, 391
08009 Barcelona
Tel: (34 3) 488-3492
Fax: (34 3) 487-7659
E-mail: barcelona@mundiprensa.es

**SRI LANKA, THE MALDIVES**
Lake House Bookshop
100, Sir Chittampalam Gardiner Mawatha
Colombo 2
Tel: (94 1) 32105
Fax: (94 1) 432104
E-mail: LHL@sri.lanka.net

**SWEDEN**
Wennergren-Williams AB
P. O. Box 1305
S-171 25 Solna
Tel: (46 8) 705-97-50
Fax: (46 8) 27-00-71
E-mail: mail@wwi.se

**SWITZERLAND**
Librairie Payot Service Institutionnel
C(tm)tes-de-Montbenon 30
1002 Lausanne
Tel: (41 21) 341-3229
Fax: (41 21) 341-3235

ADECO Van Diermen EditionsTechniques
Ch. de Lacuez 41
CH1807 Blonay
Tel: (41 21) 943 2673
Fax: (41 21) 943 3605

**THAILAND**
Central Books Distribution
306 Silom Road
Bangkok 10500
Tel: (66 2) 2336930-9
Fax: (66 2) 237-8321

**TRINIDAD & TOBAGO AND THE CARRIBBEAN**
Systematics Studies Ltd.
St. Augustine Shopping Center
Eastern Main Road, St. Augustine
Trinidad & Tobago, West Indies
Tel: (868) 645-8466
Fax: (868) 645-8467
E-mail: tobe@trinidad.net

**UGANDA**
Gustro Ltd.
PO Box 9997, Madhvani Building
Plot 16/4 Jinja Rd.
Kampala
Tel: (256 41) 251 467
Fax: (256 41) 251 468
E-mail: gus@swiftuganda.com

**UNITED KINGDOM**
Microinfo Ltd.
P.O. Box 3, Omega Park, Alton, Hampshire GU34 2PG
England
Tel: (44 1420) 86848
Fax: (44 1420) 89889
E-mail: wbank@microinfo.co.uk
URL: http://www.microinfo.co.uk

The Stationery Office
51 Nine Elms Lane
London SW8 5DR
Tel: (44 171) 873-8400
Fax: (44 171) 873-8242
URL: http://www.the-stationery-office.co.uk/

**VENEZUELA**
Tecni-Ciencia Libros, S.A.
Centro Cuidad Comercial Tamanco
Nivel C2, Caracas
Tel: (58 2) 959 5547; 5035; 0016
Fax: (58 2) 959 5636

**ZAMBIA**
University Bookshop, University of Zambia
Great East Road Campus
P.O. Box 32379
Lusaka
Tel: (260 1) 252 576
Fax: (260 1) 253 952

**ZIMBABWE**
Academic and Baobab Books (Pvt.) Ltd.
4 Conald Road, Graniteside
P.O. Box 567
Harare
Tel: 263 4 755035
Fax: 263 4 781913